ELEMENTARY NUMBER THEORY

A Collection of Problems with Solutions

V K KRISHNAN

Professor, Department of Mathematics
St. Thomas College, Thrissur, Kerala

Universities Press

ELEMENTARY NUMBER THEORY

UNIVERSITIES PRESS (INDIA) PRIVATE LIMITED

Registered Office
3-6-747/1/A & 3-6-754/1, Himayatnagar, Hyderabad 500 029 (Telangana), INDIA
e-mail: info@universitiespress.com; Web: www.universitiespress.com

Distributed by
Orient Blackswan Private Limited

Registered Office
3-6-752 Himayatnagar, Hyderabad 500 029 (Telangana), INDIA

Other Offices
Bengaluru / Bhopal / Chennai / Guwahati / Hyderabad / Jaipur
Kolkata / Lucknow / Mumbai / New Delhi / Patna / Vijayawada

First published 2012
Reprinted 2014, 2017

ISBN: 978 81 7371 743 7

Typeset by Times 10/12 by
OSDATA, Hyderabad 200 049

Printed in India at
B. B. Press
Noida

Published by
Universities Press (India) Private Limited
3-6-747/1/A & 3-6-754/1, Himayatnagar, Hyderabad 500 029 (Telangana), India
E-mail: info@universitiespress.com

Contents

Preface

This book is a collection of three hundred problems in elementary number theory together with their solutions. Before plunging into the problems, there is a brief introduction to number theory. This introductory treatment includes the basic theorems in this area. Number theory deals with the properties of integers. The most interesting and important property of integers is that of divisibility and congruence. Many of the problems in this book deal with this fascinating property.

The book is aimed at school students preparing for talent tests like the mathematical Olympiads. Most of the problems are chosen from question papers of the regional, national and international mathematical Olympiads, and the talent tests conducted by the Association of Mathematics Teachers of India. Some are taken from standard textbooks, and some are new. These problems are based on the basic properties of integers; however, the techniques used in arriving at the solution differ from one problem to another. It is difficult to arrange the problems according to the theory or the techniques. Therefore, in this book, only for the sake of getting through a fixed number of problems at a time, I have divided the three hundred problems into ten sets with thirty problems each.

The book is arranged in the following way: Chapter one introduces us to the basic properties of integers. Chapter two presents a collection of problems divided into 10 sets. The solutions of these problems are given in the next chapter. In order to make the problems as self-contained as possible, the statements of the problems are also given along with the solutions. This will make the reading easy and leisurely.

In order to make the book suitable for self-study, the proofs of theorems and the solutions to problems are given in detail and are made as self-contained as possible.

A list of books which were consulted for the preparation of this book is given at the end.

I am grateful to the reviewer for the valuable comments made after examining the manuscript. I am also grateful to the publishers for agreeing to publish this book.

V. K. Krishnan

ONE

Basic Properties of Integers

Number theory deals with the properties of integers. Positive integers, also called natural numbers, are possibly the first mathematical concept developed by humanity. The study of integers have always fascinated the human mind.

1.1 Divisibility

There are two basic operations, addition and multiplication, in the set of integers. Thus, if a and b are integers, then $a + b$ and ab are integers. Addition has an inverse operation, namely subtraction. Given integers a and b, we can find an integer c such that $a = b + c$. However, no such inverse of multiplication exists among integers; that is, given integers a and b, there need not exist an integer c such that $a = bc$. This leads to the concept of divisibility, a property which is the central theme of number theory.

Let a, b be integers and $b \neq 0$. We say that b *divides* a if there is an integer c such that $a = bc$; in this case, we write $b|a$. Clearly, $1|a$ for any a; and if $a \neq 0$, then $a|a$ and $a|0$. If b divides a, we also say that b is *divisor* of a, or a is *divisible* by b, or a is a *multiple* of b.

If $a = bc$, then $a = (-b)(-c)$, and $-a = b(-c) = (-b)c$. If c is an integer, so is $-c$. It follows that b divides a if and only if $|b|$ divides $|a|$. If $a = bc$ and c is a non-zero integer, then $|c| \geq 1$ and so $|a| = |b||c| \geq |b|$. Thus, we see that if $a|b$ and $b|a$, then $|a| \leq |b|$ and $|b| \leq |a|$, and hence, $|a| = |b|$.

The principle of mathematical induction

While dealing with the properties of multiplication and divisibility, we need to consider only positive integers; not much is gained by including negative integers. We may use $\mathbf{N}$ to denote the set of positive integers, also called *natural numbers*, and $\mathbf{Z}$ to denote the set of all integers including zero.

A property $P(n)$ concerning a variable integer n holds for all natural numbers n if it satisfies the following two conditions:

(i) $P(1)$ is true.

(ii) If $P(n)$ is true, then $P(n+1)$ is also true.

Equivalently,

(i) $P(1)$ is true.

(iia) If $P(k)$ is true for $1 \leq k \leq n$, then $P(n+1)$ is also true.

This property of natural numbers is called the principle of mathematical induction, and is one of the fundamental properties of $\mathbf{N}$.

The well-ordering principle

Every non-empty set of natural numbers has a least element. This principle is logically equivalent to the principle of induction. The set $\mathbf{N}$ has the least element 1.

1.1.1 Primes

An integer $p > 1$ is said to be a *prime* if it has no divisor other than ± 1, $\pm p$, that is, if $b|p$ and $b > 1$, then $b = p$. The first five primes are 2, 3, 5, 7 and 11. There is only one even prime, namely 2. An integer $n > 1$ is said to be *composite* if it is not prime. Note that 1 is neither a prime nor a composite; the natural numbers greater than 1 are divided into groups, namely, primes and composites.

If $n > 1$ and n is not prime, then n has a divisor greater than 1. If p is the least divisor of n and $p > 1$, then p is necessarily a prime. It follows that every integer greater than 1 is either a prime or has a prime divisor. If

p is a prime dividing n, then $n = n_1 p$, where $1 \leq n_1 < n$. If $n_1 = 1$, then n is prime. If $n_1 > 1$, we can repeat the argument with n replaced by n_1. Thus, we see that every integer greater than 1 is either a prime or a product of primes.

Theorem 1 (Division algorithm) *If a and b are integers and $b > 0$, there exist integers q and r, uniquely determined by a, b, such that*

$$a = qb + r, \quad 0 \leq r < b.$$

Proof Consider the set S of integers n expressible as $n = bx - a$ for some integer x. If $x = |a| + 1$, then

$$n = bx - a = b(|a| + 1) - a.$$

Now, $b \geq 1$ since b is an element of $\mathbf{N}$. Hence, $n \geq |a| + 1 - a \geq 1$. This shows that S contains positive integers. Let A be the set of all natural numbers in S. Then A has a least element which may be denoted by c. Then $c = bx_0 - a$ for some integer x_0, and $c \geq 1$. Thus,

$$a = bx_0 - c = b(x_0 - 1) + (b - c) = qb + r,$$

where $q = x_0 - 1$ and $r = b - c$ are integers.

Now, $c - b$ does not belong to A as $c - b < c$ and c is the least element of A. But $c - b = b(x_0 - 1) - a$ must be an element of S since $x_0 - 1$ is an integer. This implies that $c - b \leq 0$. Moreover, $b - c \leq b - 1$ as $c \geq 1$. Hence, $0 \leq r \leq b - 1$. This proves the existence of q and r.

For uniqueness, suppose that there exist integers q_1, r_1, besides q, r, such that $a = q_1 b + r_1, 0 \leq r_1 < b$. Then, $qb + r = q_1 b + r_1$, and hence,

$$(q - q_1)b = r_1 - r < b, \quad (q_1 - q)b = r - r_1 < b.$$

Since $b > 0$, this implies that $q - q_1 < 1$ and $q_1 - q < 1$. This is possible only when $q = q_1$, which in turn forces $r = r_1$. This proves the uniqueness. □

If $a = qb + r$, where q and r are integers, and $0 \leq r < b$, then q is called the *quotient* and r, the *remainder* obtained when a is divided by b. The remainder r is 0 if and only if a is divisible by b.

Let N and b be positive integers and $b > 1$. Divide N successively by b to get the quotients and remainders as follows:

$$N_i = N_{i+1}b + a_i, \quad i = 0, 1, 2, \ldots,$$

where $0 \leq a_i < b$ and $N_0 = N$. Then, $a_i b^i = N_i b^i - N_{i+1} b^{i+1}$. The quotients become zero after a certain stage. If $N_{k+1} = 0$, then

$$a_0 + a_1 b + \cdots + a_k b^k = N_0 - N_{k+1} = N.$$

This gives a representation of N in powers of b with coefficients a_i satisfying $0 \leq a_i < b$. This is the *base-b representation* of N. If $b = 10$, it is called the *decimal representation*. If $b = 2$, it is called the *binary representation*.

Suppose that N has two base-b representations as

$$N = a_0 + a_1 b + a_2 b^2 + \cdots = c_0 + c_1 b + c_2 b^2 + \cdots,$$

where $0 \leq a_i < b$ and $0 \leq c_i < b$. Divide N by b, and compare the remainders and quotients. We get

$$a_0 = c_0, \quad N_1 = a_1 + a_2 b + \cdots = c_1 + c_2 b + \cdots.$$

The process can be repeated until we find that $a_i = c_i$ for all i. Thus, the coefficients a_i in the base-b representation are uniquely determined by N and b.

1.1.2 The greatest common divisor and least common multiple

While considering divisors of an integer, we need to consider only positive divisors. If d divides a and d divides b, then we say that d is a common divisor of a and b. If $d|a$ and $d|b$, then $a = dc_1$ and $b = dc_2$, where c_1 and c_2 are integers. If x and y are integers, then

$$ax + by = dc_1 x + dc_2 y = dc,$$

where $c = c_1 x + c_2 y$, an integer. This shows that if a and b are divisible by d, then $ax + by$ is also divisible by d. This with $x = 1$ and $y = \pm 1$ shows that $a + b$ and $a - b$ are divisible by d.

If $a = qb$ and $b = q_1 c$, then $a = qq_1 c$. This shows that if $b|a$ and $c|b$, then $c|a$.

Theorem 2 *If a and b are integers of which at least one is non-zero, then there is a unique positive integer d such that*

(i) *d is a common divisor of a and b;*

(ii) *there exist integers x and y such that $d = ax + by$;*

(iii) *if c is any common divisor of a and b, then c divides d.*

Proof Let S be the set of of all positive integers n of the form $n = ax + by$, where x and y are integers. If $x = a$ and $y = b$, then $ax + by = a^2 + b^2$, which is positive. It follows that S is non-empty. Hence, S must have a least element, say d.

Since d is in S, we first observe that $d = ax + by$, where x and y are integers. To prove that d is a divisor of a, let q be the quotient and r, the remainder when a is divided by d. Then $0 \leq r < d$, and

$$r = a - qd = a - q(ax + by) = ax_1 + by_1,$$

where $x_1 = 1 - qx$ and $y_1 = -qy$ are integers. If $r \neq 0$, then $r > 0$, and so r belongs to S. Since $r < d$, this contradicts the minimality of d. It follows that $r = 0$ and $a = qd$. Thus, $d|a$. Similarly, $d|b$. Further, if c divides a and b, then it divides $ax + by = d$.

Finally, for uniqueness, suppose that d and d_1 are positive integers having the properties (i) and (iii). Since d_1 is a common divisor, we must have $d_1|d$. Since d is a common divisor, we also have $d|d_1$. This implies that $|d_1| = |d|$. Thus, $d_1 = d$, since d_1 and d are positive. This completes the proof. □

If c is any of the common divisors of a and b and d is as given by the above theorem, then $c|d$ and so $|c| \leq d$. Hence, d is the greatest of all common divisors. It is called the *greatest common divisor* (gcd) of a and b, and is denoted by $\gcd(a, b)$, or simply, by (a, b). If $\gcd(a, b) = 1$ then a and b are said to be *relatively prime* or *coprime*.

Note that any pair (a, b) of integers, of which at least one is non-zero, has a unique gcd. The gcd is always a positive integer. Moreover, $\gcd(a, 0) = |a|$ for $a \neq 0$, and

$$\gcd(a, b) = \gcd(b, a) = \gcd(a, |b|) = \gcd(|a|, |b|).$$

If p is prime, a is any integer and $d = \gcd(a, p)$, then $d = 1$ or $d = p$. Hence, if p is prime and a is not divisible by p, then $\gcd(a, p) = 1$, that is, a is relatively prime to p.

If d is a common divisor of a and b, and if $d = ax + by$ with integers x and y, then d has the properties (i), (ii) and (iii) of the theorem. Because of uniqueness, d is necessarily the gcd. Consequently, if there exist integers x and y such that $ax + by = 1$, then $\gcd(a, b) = 1$.

Let $d = \gcd(a, b)$. Then ma and mb are divisible by md for any positive integer m. Further, if $d = ax + by$, then $md = (ma)x + (mb)y$. It follows that md is the gcd of ma and mb, that is,

$$\gcd(ma, mb) = md = m\gcd(a, b) \quad \text{for } m \geq 1.$$

Consequently, if $a = a_1 d$ and $b = b_1 d$, then

$$d = \gcd(a_1 d, b_1 d) = d\gcd(a_1, b_1),$$

and hence, $\gcd(a_1, b_1) = 1$. In other words, if $d = \gcd(a, b)$, then a/d and b/d are coprime.

Suppose that a and b are positive integers and $\gcd(a, b) = d$. Then $d = ax + by$ with integers x and y. Hence,

$$d = a(x + rb) - b(-y + ra), \quad \text{for all integers } r.$$

If r is large enough, then both $rb + x$ and $ra - y$ are positive integers. Thus, we see that if a and b are positive integers, then the gcd can be expressed as $d = au - bv$, where u and v are positive integers.
Let a be relatively prime to b and c. Then

$$ax + by = 1 \quad \text{and} \quad ax_1 + cy_1 = 1,$$

with integers x, y, x_1, y_1. On multiplication,

$$1 = a^2 x x_1 + acxy_1 + abx_1 y + bcyy_1 = ax_2 + bcy_2,$$

where $x_2 = axx_1 + cxy_1 + bx_1 y$ and $y_2 = yy_1$. This shows that a is relatively prime to bc. On the other hand, if $ax_2 + bcy_2 = 1$, then $ax_2 + bu = 1$ and $ax_2 + cv = 1$. Thus, we see that a is relatively prime to bc if and only if a is relatively prime to both b and c.

More generally, if a is relatively prime to n integers $b_1, \ldots, b_n$, then it is relatively prime to their product $b_1 \cdots b_n$. This has been verified for $n = 2$. Assume that it is true for $n-1$ integers $b_1, \ldots, b_{n-1}$. Then, a is relatively prime to their product $b_1 \cdots b_{n-1}$. Use the special case proved above with $b = b_1 \cdots b_{n-1}$ and $c = b_n$. This proves, by the principle of mathematical induction, that the general result holds for all $n \geq 2$. In particular, if $\gcd(a, b) = 1$, then $\gcd(a, b^n) = 1$ for all positive integers n. Since $(a, b) = (b, a)$, we also get

$$\gcd(a, b) = 1 \Longrightarrow \gcd(a^m, b^n) = 1 \quad \text{for all positive integers } m, n.$$

Let $a + bx = c$, where a, b, c and x are integers. If $b \neq 0$, let $d_1 = \gcd(a, b)$ and $d_2 = \gcd(b, c)$. Then, $d_1 | c$, and hence $d_1 | d_2$. Since $a = c - bx$, we can also see that $d_2 | d_1$. It follows that $d_1 = d_2$, that is, $\gcd(a, b) = \gcd(b, c)$. This idea can be used to evaluate the gcd of two positive integers. Given a and b, divide a by b. If q is the quotient and r is the remainder, then $a = qb + r$, $0 \leq r < b$. So $\gcd(a, b) = \gcd(b, r)$. If $r \neq 0$, then $0 < r < b$; divide b by r. We have $b = q_1 r + r_1$, where $0 \leq r_1 < r$; we get $\gcd(b, r) = \gcd(r, r_1)$. If $r_1 \neq 0$, repeat this process. In general, if $r_i \neq 0$, $i \geq -1$, let

$$r_{i-1} = q_{i+1} r_i + r_{r+1}, \quad 0 \leq r_{i+1} < r_i,$$

where we take $r_{-2} = a$, $r_{-1} = b$, $r_0 = r$ and $q_0 = q$. The process can be continued as long as $r_i \neq 0$. If $r_k \neq 0$ and $r_{k+1} = 0$, then we have

$$\gcd(r_{i-1}, r_i) = \gcd(r_i, r_{i+1}), \quad \text{for } 0 \leq i \leq k.$$

It follows that

$$\gcd(a, b) = \gcd(b, r) = \gcd(r_k, 0) = r_k.$$

This procedure for evaluating $\gcd(a, b)$ is called the *Euclidean algorithm*.

Let a, b be positive integers, $d = \gcd(a, b)$ and $m = ab/d$. Then, $a = a_1 d$, $b = b_1 d$ and

$$m = a_1 b_1 d = a_1 b = a b_1.$$

This shows that m is a multiple of a and b. Further, if n is any common multiple of a and b, then

$$\frac{n}{m} = \frac{nd}{ab} = \frac{n(ax + by)}{ab} = \frac{n}{b}x + \frac{n}{a}y,$$

where x and y are integers. This shows that the positive integer m has the properties

(i) m is a common multiple of a and b, and

(ii) if n is any common multiple of a and b, then n is also a multiple of m.

The number m is called the *least common multiple* (lcm) of a and b, and is denoted by $\text{lcm}[a, b]$, or simply, by $[a, b]$.

As for gcd, we have for $k \geq 1$,

$$\text{lcm}[ka, kb] = \frac{(ka)(kb)}{\gcd(ka, kb)} = k\frac{ab}{\gcd(a, b)} = k\,\text{lcm}[a, b].$$

The concept of gcd and lcm can be extended to more than two numbers. If a, b, c are integers, and a or b is non-zero, let $d_1 = \gcd(a, b)$. Then, $d_1 > 0$. Let $d = \gcd(d_1, c)$. Then, $d_1|a$ and $d_1|b$. Since $d|d_1$, we get $d|a$ and $d|b$, besides $d|c$. Further, $d_1 = ax + by$, and hence,

$$d = d_1x_1 + cy_1 = axx_1 + byx_1 + cy_1 = ax_2 + by_2 + cy_1,$$

where x, y, x_i, y_i are integers. This shows that d is a common divisor of a, b and c, and is expressible as a linear combination of these three. Any common divisor of a, b and c is also a divisor of d. Such a positive integer d is uniquely determined. The integer d is denoted by $\gcd(a, b, c)$ and is called the gcd of a, b and c. Thus,

$$\gcd(a, b, c) = \gcd((a, b), c).$$

Similarly, if a, b and c are positive integers, let $m = \text{lcm}[[a, b], c]$. Then, m is a common multiple of a, b and c, and is a multiple of all other such multiples. We denote m by $\text{lcm}[a, b, c]$.

More generally, if $a_1, \ldots, a_n$ are positive integers, we can define $b_1, \ldots, b_n$ as follows:

$$b_1 = a_1, \quad b_i = \gcd(b_{i-1}, a_i), \quad \text{for } i \geq 2.$$

Let $d = b_n$. The positive integer d is common divisor for $a_1, \ldots, a_n$, and is expressible as

$$d = a_1x_1 + \cdots + a_nx_n \quad \text{with integers } x_i.$$

Moreover, d is divisible by all such common divisors. It is denoted by $\gcd(a, \ldots, a_n)$.

Let $c_1, \ldots, c_n$ be defined by

$$c_1 = a_1, \quad c_i = \text{lcm}[c_{i-1}, a_i], \quad \text{for } i \geq 2.$$

Then, $m = c_n$ is a common multiple of a_i, and is a divisor of all such common multiples. Denote m by $\text{lcm}[a_1, \ldots, a_n]$.

Theorem 3 (Euclid's lemma) *If a divides bc and a is relatively prime to b, then a divides c.*

Proof If a divides bc, then $bc = ma$, for some integer m. If a is relatively prime to b, then there exist integers x and y such that $ax+by = 1$. Hence,

$$c = acx + bcy = acx + may = a(cx + by).$$

Since $cx + by$ is an integer, this shows that $a|c$. □

Corollary 3.1 *If c is divisible by a and b, where a and b are coprime, then c is divisible by the product ab. More generally, if c is divisible by each of $a_1, \ldots, a_n$, which are pair-wise coprime, then c is divisible by their product $a_1 \cdots a_n$.*

Proof If c is divisible by a, then $c = ak$ for some integer k. If c is also divisible by b, then $b|ak$. Since $\gcd(b, a) = 1$, we get $b|k$, and so $k = bk_1$. Hence, $c = abk_1$. This proves the special case. The general result follows by induction. □

Corollary 3.2 *If a prime divides a product, then it divides at least one of the factors.*

Proof Let p be prime and $p|ab$. If $d = \gcd(p, a)$, then $d = 1$ or p. If $d = p$, then $p|a$. If $d = 1$, then p is relatively prime to a, and hence $p|b$. Thus, p divides a or b. Using the principle of induction, we see that if p divides $a_1 \cdots a_n$, then $p|a_i$ for some i. □

We observed earlier that every integer greater than 1 is a product of primes with the understanding that the 'product' may consist of just one factor. In fact, such an expression is unique except for the order of the factors.

Theorem 4 (The fundamental theorem of arithmetic) *An integer $n > 1$ can be expressed as a product of primes in only one way in the sense that if*

$$n = p_1 \cdots p_k \quad \text{and} \quad n = q_1 \cdots q_r,$$

where p_i and q_j are primes, not necessarily distinct, then $k = r$, and each p_i equals some q_j and each q_j equals some p_i.

Proof We shall prove by induction on k that if an integer n can be represented as a product of k primes, then any such representation has the same prime factors, possibly written in a different order. If $k = 1$, then n is prime. By the definition of prime, n cannot be expressed as a product of primes in any other way; and so the result holds in this case.

Assume that if $k > 1$, then the result is true for integers expressible as a product of $k-1$ primes. Suppose that n has two expressions as given in the theorem. The prime p_1 divides n, and so it divides the product $q_1 \cdots q_r$. By the corollary of Euclid's lemma, we see that p_1 divides some q_j. Let $p_1 | q_t$. Then, $p_1 = q_t$ by the definition of prime. Cancelling p_1 and q_t from both sides, we see that $n_1 = n/p_1$ is the product of $k-1$ primes $p_2, \ldots, p_k$ and also as the product of $r-1$ primes $q_j, j \neq t$. By our induction hypothesis, these $r - 1$ primes q_j, other than q_t, are the same as the $k - 1$ primes $p_2, \ldots, p_k$, possibly written in a different order. It follows that the r primes q_j, including q_t, are the same as the k primes $p_1, \ldots, p_k$. This completes the induction proof. □

Combining equal factors, we can express an integer $n > 1$ as

$$n = p_1^{n_1} \cdots p_k^{n_k},$$

where the p_i are distinct primes dividing n and the n_i are positive integers. Such a representation is unique, and is called the *standard representation* or *canonical representation* of n. If convenient, we may include some primes p_i which are not actually divisors of n by allowing the corresponding n_i to be 0.

The gcd and lcm of two or more positive integers can determined using the standard representations. Let

$$a = p_1^{a_1} \cdots p_k^{a_k}, \quad b = p_1^{b_1} \cdots p_k^{b_k},$$

where p_i are distinct primes and $a_i \geq 0$, $b_i \geq 0$. Then,

$$\gcd(a, b) = p_1^{r_1} \cdots p_k^{r_k}, \quad \text{lcm}[a, b] = p_1^{s_1} \cdots p_k^{s_k},$$

where $r_i = \min(a_i, b_i)$ and $s_i = \max(a_i, b_i)$. To see this, note that the gcd is a common divisor and is a multiple of all common divisors, and the lcm is a common multiple and is a divisor of all common multiples. If, further,

$$c = p_1^{c_1} \cdots p_k^{c_k}, \quad c_i \geq 0,$$

then

$$\gcd(a, b, c) = p_1^{r_1} \cdots p_k^{r_k}, \quad \text{lcm}[a, b, c] = p_1^{s_1} \cdots p_k^{s_k},$$

where $r_i = \min(a_i, b_i, c_i)$ and $s_i = \max(a_i, b_i, c_i)$.

There are an infinite number of primes. If this is not true, suppose that

$$S = \{p_1, p_2, \ldots, p_n\}$$

is the set of all primes, and let P be their product, $p_1 \cdots p_n$. Then, P is divisible by each of these primes p_i. Hence, $N = P+1$ cannot be divisible by any prime in S. But N must have a prime divisor as it is greater than 1. Consequently, there must exist primes not belonging to the set S. This is true for any finite set S of primes. It follows that there exists an infinite number of primes.

Let p_n denote the nth prime so that $p_1 = 2, p_2 = 3, p_3 = 5$, and so on. The integer $N = p_1 \cdots p_n + 1$ constructed above must have a prime divisor, which must be greater than p_n. It follows that

$$p_{n+1} \leq p_1 p_2 \cdots p_n + 1.$$

This is only a crude estimate. In fact, if n is any integer greater than 1, then there is a prime p with $n < p < 2n$. This result is known as *Bertrand's theorem*; the proof of this is beyond the scope of this book. Using this, we can prove that if p_n is the nth prime, then $p_n < 2^n$ for all $n \geq 2$. Since $p_2 = 3 < 2^2$, this holds when $n = 2$. By Bertrand's theorem, there is a prime p with $p_n < p < 2p_n$. Consequently, if $p_n < 2^n$, then

$$p_{n+1} \leq p < 2p_n < 2^{n+1}.$$

It follows by induction that $p_n < 2^n$ for all $n \geq 2$.

1.1.3 The binomial coefficients

If n is positive integer, the product of the first n natural numbers $1, 2, \ldots, n$ is denoted by $n!$. Besides, we define $0! = 1$. The *binomial coefficients* $\binom{n}{r}$ are defined by

$$\binom{n}{r} = \frac{n(n-1)\cdots(n-r+1)}{r(r-1)\cdots 1} = \frac{n!}{r!(n-r)!}, \quad \text{for } 1 \leq r \leq n,$$

and $\binom{n}{0} = 1$.

Theorem 5 *For any positive integer n, the binomial coefficients $\binom{n}{r}$ are positive integers for $0 \leq r \leq n$.*

Proof If $n = 1$, then the binomial coefficients are 1, 1. If $n = 2$, they are 1, 2, 1. These are integers. Assume that, for some n, the numbers $\binom{n}{r}$ are integers for $1 \leq r \leq n$. We have

$$\binom{n}{r} + \binom{n}{r-1} = \frac{n(n-1)\cdots(n-r+2)}{r!}\{(n-r+1)+r\}$$

$$= \frac{(n+1)n\cdots(n-r+2)}{r!} = \binom{n+1}{r},$$

for $1 \leq r \leq n$. This implies that $\binom{n+1}{r}$ is an integer for $1 \leq r \leq n$. Since

$$\binom{n+1}{0} = 1 = \binom{n+1}{n+1},$$

the theorem is proved by induction. □

Corollary 5.1 *The product of r consecutive positive integers is divisible by $r!$.*

Proof If n is any positive integer, then

$$n(n+1)\cdots(n+r-1) = \binom{n+r-1}{r} r!$$

is divisible by $r!$. □

Corollary 5.2 *If p is prime, then the binomial coefficients $\binom{p}{r}$ are divisible by p for $1 \leq r \leq p-1$.*

Proof If $1 \leq r \leq p-1$, then p is relatively prime to each of the integers $1, \ldots, r$, and hence to $r!$. Since $\binom{p}{r}$ is an integer, we see that $r\,!$ divides

$$\binom{p}{r} r! = p(p-1)\cdots(p-r+1) = pN,$$

where N is an integer. Hence, $r!$ divides N by Euclid's lemma. If $N = kr!$, then $\binom{p}{r} = pk$. □

1.1.4 Linear Diophantine equations

Polynomial equations in two or more unknowns, whose solutions are required to be integers, are known, for historical reasons, as *Diophantine equations*. Solutions of linear Diophantine equations in two unknowns have been studied in detail.

Theorem 6 *Let a, b be non-zero integers and $\gcd(a, b) = d$. Then the linear Diophantine equation $ax + by = c$ has a solution if and only if $d|c$. If $x = x_0, y = y_0$ is any particular solution, then all other solutions are given by*

$$x = x_0 + \frac{b}{d}t, \quad y = y_0 - \frac{a}{d}t,$$

where t is an arbitrary integer.

Proof If x_0 and y_0 are integers such that $ax_0 + by_0 = c$, then $d|c$ as d divides both a and b. On the other hand, there exist integers x_1, y_1 such that $ax_1 + by_1 = d$. If $c = c_1 d$, then

$$a(c_1 x_1) + b(c_1 y_1) = c_1 d = c,$$

and so the equation has a solution $x = c_1 x_1$, $y = c_1 y_1$.

Finally, suppose that x, y and x_0, y_0 are solutions in integers. Then, $ax + by = c = ax_0 + by_0$. This can be written as

$$a(x - x_0) = b(y_0 - y), \quad a_1(x - x_0) = b_1(y_0 - y),$$

where $a_1 = a/d$ and $b_1 = b/d$ are non-zero integers. Since a_1 is relatively prime to b_1, this implies that a_1 divides $y_0 - y$. If $y_0 - y = a_1 t$ for some integer t, then $x - x_0 = b_1 t$, and hence,

$$x = x_0 + b_1 t, \quad y = y_0 - a_1 t,$$

as desired. □

1.2 Congruences

We present the idea of divisibility in a different notation.

Definition Let m be a positive integer. If a and b are integers such that $a-b$ is divisible by m, then we say that a is congruent to b modulo m, and write $a \equiv b \pmod{m}$. The integer m is called *modulus*.

Let r_1 and r_2 be the remainders obtained when two integers a and b are divided by m. Then,

$$a = q_1 m + r_1, \quad b = q_2 m + r_2, \quad \text{where } 0 \le r_1, r_2 \le m-1.$$

Hence, $a - b = (q_1 - q_2)m + (r_1 - r_2)$ is divisible by m if and only if $r_1 - r_2$ is divisible by m. But $|r_1 - r_2| \le m - 1$. So $r_1 - r_2$ is divisible by m if and only if $r_1 - r_2 = 0$. Thus, we see that $a \equiv b \pmod{m}$ if and only if the remainders r_1 and r_2 are equal.

For a fixed modulus m we have a relation '$\equiv$' which behaves almost like '='. It is an *equivalence* relation, that is, it is reflexive, symmetric and transitive:

(i) *Reflexivity.* $a \equiv a$ for all a.

(ii) *Symmetry.* $a \equiv b \Longrightarrow b \equiv a$.

(iii) *Transitivity.* $a \equiv b$ and $b \equiv c \Longrightarrow a \equiv c$.

The congruence relation has some more properties. If $a \equiv b \pmod{m}$, then $ac \equiv bc \pmod{m}$ for all integers c. This follows from the fact that if $b - a$ is divisible by m, so is $bc - ac$. Let $a \equiv b \pmod{m}$ and $c \equiv d \pmod{m}$. Then, $(b-a)+(d-c)$ is divisible by m, and hence, $a+c \equiv b+d \pmod{m}$. Further,

$$ac \equiv bc \pmod{m}, \quad bc \equiv bd \pmod{m},$$

and hence, $ac \equiv bd \pmod{m}$. Thus, we see that congruences can be added and multiplied. It follows that if $a \equiv b \pmod{m}$, then $a^n \equiv b^n \pmod{m}$ and $ka^n \equiv kb^n \pmod{m}$ for all integers k and positive integers n. More generally, if $a \equiv b \pmod{m}$, and if $f(x)$ is a polynomial with integer coefficients, then $f(a) \equiv f(b) \pmod{m}$.

Suppose that

$$a \equiv b \pmod{m_i}, \quad \text{for } i = 1, \ldots, k.$$

Then, $a-b$ is divisible by each m_i. If the moduli m_i are pair-wise coprime, then $a - b$ is divisible by their product $m_1 \cdots m_k$, by the corollary of Euclid's lemma, that is,

$$a \equiv b \pmod{m_1 \cdots m_k}.$$

If $ac = bc$ and $c \neq 0$, then the common factor can be cancelled from both sides. However, in a congruence relation, the cancellation is valid only when the factor to be cancelled is relatively prime to the modulus.

Theorem 7 (The cancellation law) *If $ac \equiv bc \pmod{m}$ and c is relatively prime to m, then $a \equiv b \pmod{m}$.*

Proof If $ac \equiv bc \pmod{m}$, then $c(b-a) = cb - ca$ is divisible by m. If, further, m is relatively prime to c, then m divides the other factor $b - a$ by Euclid's lemma. □

Corollary 7.1 *If $ab \equiv cd \pmod{m}$ and $b \equiv d \pmod{m}$, with $\gcd(b, m) = 1$, then $a \equiv c \pmod{m}$.*

Proof If $b \equiv d \pmod{m}$, then $cb \equiv cd \pmod{m}$. If, further, $cd \equiv ab \pmod{m}$, then $cb \equiv ab \pmod{m}$. We may cancel b as it is relatively prime to m. □

If $a \equiv b \pmod{m}$, then $a = km + b$ for some integer k, and so $\gcd(a, m) = \gcd(m, b)$, as observed earlier. Thus, if two integers are congruent modulo m, then one of them is relatively prime to m if and only if the other is relatively prime to it.

1.2.1 Residue systems

Consider a fixed modulus m. For $r = 0, 1, \ldots, m-1$, let A_r denote the set of all integers which give the same remainder r when divided by m. Any two integers in the same set A_r are congruent modulo m; two integers chosen from different sets A_i and A_j are incongruent modulo m. Thus, we get a partitioning of the set of all integers into m disjoint classes $A_0, A_1, \ldots, A_{m-1}$. These are called *residue classes* modulo m. Choosing one element from each class A_i, we get a set S of m elements with the following property: Every integer n is congruent modulo m to one and only one element in S. A set S of m integers with this property is called a *complete residue system* modulo m.

Suppose that we have a set of m integers no two of which are congruent modulo m. Two integers belong to the same residue class if and only if they are congruent modulo m. It follows that any set $\{a_1, \ldots, a_m\}$ of m integers form a complete residue system modulo m if and only if no two of them are congruent modulo m. The sets $\{0, 1, \ldots, m-1\}$, $\{1, 2, \ldots, m\}$, and more generally, $\{k, k+1, \ldots, k+m-1\}$, where k is any integer, are complete residue systems modulo m.

1.2.2 Linear congruences

Let m be a fixed modulus, and $f(x)$ be a polynomial with integer coefficients. The congruence $f(x) \equiv 0 \pmod{m}$ is called a polynomial congruence. Here the variable x is supposed to take integer values only.

If x_1 is an integer satisfying the congruence $f(x) \equiv 0 \pmod{m}$ and x_2 is any integer congruent modulo m to it, then $f(x_2) \equiv f(x_1) \equiv 0 \pmod{m}$. Hence, if x_1 is a root of the congruence, so is x_2. Consequently, while counting the number of roots of a polynomial congruence, we need not distinguish between congruent roots. The number of roots is the number of integers satisfying the congruence and belonging to a complete residue system.

Theorem 8 *If a is relatively prime to m, then the congruence $ax \equiv b \pmod{m}$ has a solution. The solution is unique modulo m, that is, if x_1 and x_2 are solutions, then $x_1 \equiv x_2 \pmod{m}$.*

Proof Since a is relatively prime to m there exist integers u and v such that $au + mv = 1$. Hence,

$$au = 1 - mv \equiv 1 \pmod{m}, \quad a(bu) \equiv b \pmod{m}.$$

This gives a solution $x = bu$.
For uniqueness, suppose that

$$ax_1 \equiv b \pmod{m} \quad \text{and} \quad ax_2 \equiv b \pmod{m}.$$

Then, $a(x_1 - x_2) \equiv 0 \pmod{m}$, and so $x_1 \equiv x_2 \pmod{m}$ by cancellation law. □

If x_1 is a solution of the congruence, then any other solution is congruent to x_1; that is, the general solution is of the form $x_1 + mn$ with integers n.

Theorem 9 (Chinese remainder theorem) *Let $m_1, m_2, \ldots, m_k$ be positive integers, any two of which are relatively prime. Then the system of congruences*

$$x \equiv b_i \pmod{m_i}, \quad i = 1, 2, \ldots, k,$$

has a solution. The solution is unique modulo the product $m_1 m_2 \cdots m_k$ of the moduli.

Proof Let M be the product of all the moduli, and $M_i = M/m_i$. Then M_i is the product of all m_j with $j \neq i$. Since all m_i are pair-wise coprime, we see that m_i is relatively prime to M_i, and hence, $M_i u + m_i v = 1$ for some integers u and v. Denote this u by M_i'. Then, $M_i M_i' \equiv 1 \pmod{m_i}$. Since M_i is divisible by m_j for $j \neq i$, we see that $M_i M_i' \equiv 0 \pmod{m_j}$ for $j \neq i$. Let

$$x = b_1 M_1 M_1' + b_2 M_2 M_2' + \cdots + b_k M_k M_k'.$$

Then, $x \equiv b_i \pmod{m_i}$. This holds for each i. Thus, we get a solution.

For uniqueness, suppose that x and y satisfy all the k congruences simultaneously. Then,

$$x \equiv b_i \equiv y \pmod{m_i}, \quad x - y \equiv 0 \pmod{m_i},$$

for $i = 1, 2, \ldots, k$. Since the m_is are pair-wise coprime, $x - y$ is divisible by their product $m_1 m_2 \cdots m_k = M$. Hence, $x \equiv y \pmod{M}$. This completes the proof. □

The above proof of the existence of a solution can be illustrated by an example. Consider the system of three congruences

$$x \equiv 2 \pmod 3, \quad x \equiv -12 \pmod 7, \quad x \equiv 4 \pmod{13}.$$

In the notation of the proof we have

$$M = 3 \cdot 7 \cdot 13 = 273, \quad M_1 = 91, \quad M_2 = 39, \quad M_3 = 21.$$

We have to choose M_i' such that $M_i M_i' \equiv 1 \pmod{m_i}$. Since

$$M_1 \equiv 1 \pmod 3, \quad M_2 \equiv 4 \pmod 7, \quad M_3 \equiv 8 \pmod{13},$$

these are satisfied by $M_1' = 1$, $M_2' = 2$ and $M_3' = 5$. So let

$$x = 2 \cdot 91 \cdot 1 + (-12) \cdot 39 \cdot 2 + 4 \cdot 21 \cdot 5 = -334.$$

So -334 is a solution. The general solution is $-334 + 273n$ for integers n. The least positive solution is 212.

To establish the existence of a solution, we can also proceed as follows: The general solution of the first congruence is $x = b_1 + m_1 n$. This will also satisfy the second if the integer n is chosen properly. The procedure is illustrated by the same system of congruences considered above. The first gives $x = 3n + 2$, and then the second becomes $3n + 2 \equiv -12 \equiv 2 \pmod 7$. This implies that $n \equiv 0 \pmod 7$, and hence

$$n = 7k, \quad x = 3(7k) + 2.$$

Then the third becomes

$$21k + 2 \equiv 4 \pmod{13},$$

giving $k \equiv 10 \pmod{13}$, $k = 13r + 10$. Hence, $x = 21(13r + 10) + 2 = 273r + 212$. The least positive solution is 212.

Theorem 10 (Wilson's theorem) *If p is prime, then* $(p-1)! + 1 \equiv 0 \pmod p$.

Proof If $p = 2$ or 3, then $(p-1)! + 1 = 2$ or 3, respectively. So let $p \geq 5$. If a is any of the integers $1, 2, \ldots, p-1$, then a is relatively prime to p, and hence, the congruence $ax \equiv 1 \pmod p$ has a solution unique modulo p. This means that each a in the set $\{1, 2, \ldots, p-1\}$ has an *associate* a', in the same set, such that $aa' \equiv 1 \pmod p$.

Suppose that the associate a' of some a is the same as a itself. Then $a^2 \equiv 1 \pmod p$; and hence, p divides either $a-1$ or $a+1$, which holds only when $a = 1$ or $p-1$. For $2 \leq a \leq p-2$, the associate a' is different from a; and moreover, $2 \leq a' \leq p-2$. Consider the product of the $p-3$ numbers $2, \ldots, p-2$. The factors can be grouped into $(p-3)/2$ pairs (a, a'). The product of each group is congruent to 1 modulo p. It follows that

$$2 \cdot 3 \cdots (p-2) \equiv (1)(1) \cdots (1) \pmod p.$$

Since $p-1 \equiv -1 \pmod p$, we get $(p-1)! \equiv -1 \pmod p$, which proves the theorem. □

Wilson's theorem has a converse which states that if

$$(n-1)! + 1 \equiv 0 \pmod n \quad \text{and} \quad n > 1,$$

then n is prime. To see this, suppose that n is composite and $n = ab$, where $2 \leq a \leq n-1$. Then a divides $(n-1)!$. In this case, $(n-1)!+1$ cannot be divisible by a, and hence it cannot be divisible by n.

1.2.3 Lagrange's theorem

Let $f(x)$ be a polynomial with integer coefficients. We observed earlier that if x_1 is a solution of the polynomial congruence $f(x) \equiv 0 \pmod{m}$ and $x_1 \equiv x_2 \pmod{m}$, then x_2 is also a solution. The number of roots is the number of integers in a complete residue system, satisfying the congruence.

Theorem 11 (Lagrange's theorem) *Let $f(x) = a_0x^n + a_1x^{n-1} + \cdots + a_n$, where the coefficients a_i are integers. Let p be a prime. If the congruence $f(x) \equiv 0 \pmod{p}$ is satisfied by more than n values of x, which are incongruent modulo p, then each a_i is divisible by p.*

Proof Suppose that $x_0, x_1, \ldots, x_n$ are $n+1$ integers, incongruent modulo p which satisfy the congruence. We shall prove, by induction on n, that each a_i is divisible by p. If $n = 1$, then we have

$$a_0x_0 + a_1 \equiv 0 \pmod{p}, \quad a_0x_1 + a_1 \equiv 0 \pmod{p},$$

and so $a_0(x_0 - x_1) \equiv 0 \pmod{p}$. Since $x_0 - x_1$ is not divisible by p, this implies that a_0 is divisible by p, and hence, a_1 is divisible by p.

Let $n > 1$, and assume that the result holds for all polynomials of degree less than n. We have

$$f(x) - f(x_0) = \sum_{i=0}^{n-1} a_i(x^{n-i} - x_0^{n-i}) = (x - x_0)g(x),$$

where $g(x)$ is a polynomial with the same leading coefficient a_0 and is of degree $n-1$. Since

$$(x_i - x_0)g(x_i) = f(x_i) - f(x_0) \equiv 0 \pmod{p} \quad \text{for all } i \geq 1,$$

we see that $g(x_i) \equiv 0 \pmod{p}$ for $1 \leq i \leq n$. Thus, the congruence $g(x) \equiv 0 \pmod{p}$ is satisfied by n incongruent values of x. Note that the coefficients of $g(x)$ are integers. Hence, the coefficients of $g(x)$ are divisible by p. In particular, a_0 is divisible by p.

Let $f_1(x) = f(x) - a_0x^n = a_1x^{n-1} + \cdots + a_n$. Then the congruence $f_1(x) \equiv 0 \pmod{p}$ is satisfied by $n+1$ incongruent values of x. By induction hypothesis, the coefficients of $f_1(x)$ are divisible by p; that is, a_i is divisible by p for $i \geq 1$. This completes the proof. □

1.3 Fermat's Theorem

Fermat's theorem, often called *Fermat's little theorem,* is one of the most powerful theorems in elementary number theory. It is called *little* to distinguish it from *Fermat's last theorem,* which remained an unproved conjecture for more than three centuries after his death. We consider only the little one.

Theorem 12 (Fermat's theorem) *If* p *is prime, then* $a^{p-1} \equiv 1 \pmod{p}$ *for all integers* a *which are not divisible by* p*, and equivalently,* $a^p \equiv a \pmod{p}$ *for all integers* a.

Proof For $k = 1, 2, \ldots, p-1$, let r_k be the remainder when ka is divided by p. Then,

$$ka \equiv r_k \pmod{p}, \quad 0 \leq r_k \leq p-1.$$

Multiplying these congruences for $k = 1, 2, \ldots, p-1$, we get

$$(p-1)!\, a^{p-1} \equiv r_1 \cdot r_2 \cdots r_{p-1} \pmod{p}. \tag{*}$$

Suppose that a is not divisible by p. Then it is relatively prime to p, and hence,

$$r_k = 0 \Longrightarrow ka \equiv 0 \pmod{p} \Longrightarrow k \equiv 0 \pmod{p}.$$

This shows that $r_k \neq 0$ for $1 \leq k \leq p-1$. Further,

$$r_j = r_k \Longrightarrow ja \equiv ka \pmod{p} \Longrightarrow j \equiv k \pmod{p}.$$

But $|j-k| < p$ for $1 \leq j, k \leq p-1$ and so $r_j = r_k$ only when $j = k$. It follows that the $p-1$ remainders $r_1, \ldots, r_{p-1}$ are distinct, and none of them is 0. So they are precisely the numbers $1, 2, \ldots, p-1$, written in some order, and hence their product is $(p-1)!$. It follows from (*) that

$$(p-1)!\, a^{p-1} \equiv (p-1)! \pmod{p}.$$

Since $(p-1)!$ is relatively prime to p, we can cancel it and get $a^{p-1} \equiv 1 \pmod{p}$. This proves the first statement. To see the equivalence of the two statements, note that $a^p - a = a\left(a^{p-1} - 1\right)$ is divisible by p if and only if a or $a^{p-1} - 1$ is divisible by p. This completes the proof. □

Another proof: Since the binomial coefficient $\binom{p}{r}$ is divisible by p for $1 \le r \le p-1$, we see from the binomial expansion of $(k+1)^p$ that

$$(k+1)^p - k^p \equiv 1 \pmod{p}.$$

Using this with $k = 1, 2, \ldots, n-1$ we get, on addition,

$$n^p - 1 \equiv n - 1 \pmod{p}, \quad n^p \equiv n \pmod{p}.$$

This holds for $n \ge 2$, and is trivial when $n = 0$ or 1.

If $n < 0$, we get $(-n)^p \equiv -n \pmod{p}$ by what has been proved above. If p is odd, then $(-n)^p = -n^p$. Since $(-n)^2 = n^2 \equiv -n^2 \pmod{2}$, we see that

$$-n^p \equiv (-n)^p \equiv -n \pmod{p}.$$

Hence $n^p \equiv n \pmod{p}$ for all integers n and all primes p. □

Corollary 12.1 *If p is an odd prime, then all the coefficients in the polynomial*

$$f(x) = (x-1)(x-2)\cdots(x-p+1) - x^{p-1} + 1$$

are divisible by p.

Proof Let $g(x) = (x-1)(x-2)\cdots(x-p+1)$ and $h(x) = x^{p-1} - 1$. Then, $h(x) \equiv 0 \pmod{p}$ for $x = 1, 2, \ldots, p-1$ by Fermat's theorem, and $g(x) \equiv 0 \pmod{p}$ for these values of x. Hence, $f(x) \equiv 0 \pmod{p}$ has $p-1$ roots incongruent modulo p. But $f(x) = g(x) - h(x)$ is of degree less than $p-1$. So the result follows from Theorem ??. □

If p is an odd prime, then the constant term in the polynomial $f(x)$ of this corollary is $(-1)^{p-1}(p-1)! + 1 = (p-1)! + 1$. It follows that Wilson's theorem is a special case of this corollary.

1.3.1 Pseudoprimes and Carmichael numbers

By Fermat's theorem, $2^n \equiv 2 \pmod{n}$ if n is prime. There are, however, composite numbers n such that $2^n \equiv 2 \pmod{n}$. For example, consider the number $n = 341 = 11 \times 31$. Fermat's theorem with $p = 11$ shows that $2^{10} \equiv 1 \pmod{11}$, and hence, $2^{340} \equiv 1 \pmod{11}$. Since $2^5 = 32 \equiv 1 \pmod{31}$, we get $2^{340} \equiv 1 \pmod{31}$. Consequently, $2^{340} - 1$ is divisible by 11 and 31, and hence, by $11 \times 31 = n$. Thus, $2^n \equiv 2 \pmod{n}$. In fact, 341 is the smallest composite number with this property. A composite number n is said to be a *pseudoprime* if $2^n - 2$ is divisible by n.

Fermat's theorem asserts that if n is prime and a is relatively prime to n, then $a^{n-1} \equiv 1 \pmod{n}$. There are, in fact, composite numbers n with this property. For example, consider the number $n = 561 = 3 \cdot 11 \cdot 17$. If a is relatively prime to n, then a is not divisible by 3, 11 or 17. By Fermat's theorem with these primes, we get

$$a^2 \equiv 1 \pmod{3}, \quad a^{10} \equiv 1 \pmod{11}, \quad a^{16} \equiv 1 \pmod{17}.$$

Since 560 is divisible by 2, 10 and 16, we see that a^{560} is congruent to 1 modulo 3, 11 and 17. Hence, $a^{560} - 1$ is divisible by $3 \cdot 11 \cdot 17 = n$.

A composite number n is said to be a *Carmichael number* if $a^{n-1} \equiv 1 \pmod{n}$ for all natural numbers a which are relatively prime to n. A Carmichael number can also be defined as a composite number n such that $a^n \equiv a \pmod{n}$ for all natural numbers a. It will be shown in Chapter 3 that these two conditions are equivalent. The smallest Carmichael number is the number 561 considered above.

The pseudoprimes less than 2000 are 341, 561, 645, 1105, 1387, 1729 and 1905. Of these, only 561, 1105 and 1729 are Carmichael numbers. Carmichael numbers are odd, while a pseudoprime can be even.

1.4 Number-Theoretic Functions

A function $f(n)$ defined for natural numbers n is often called, in number theory, a *number-theoretic* (or *arithmetic*) *function*. Thus, an arithmetic function is nothing but a sequence. We shall consider some such elementary functions.

1.4.1 Euler's function

The arithmetic function $\phi(n)$, called *Euler's function*, is defined as follows.

Definition If n is a positive integer, then the number of positive integers not exceeding n and relatively prime to n is denoted by $\phi(n)$.

Thus $\phi(1) = 1$, and if $n > 1$, then $\phi(n)$ is the number of positive integers less than n and relatively prime to n. Thus, $\phi(n) \le n - 1$ for all $n > 1$. If p is prime, then all the integers $1, 2, \ldots, p - 1$ are relatively prime to p. Thus, $\phi(p) = p - 1$, if p is prime. For $n = 1, 2, 3, 4, 5, 6$, the values of $\phi(n)$ are 1, 1, 2, 2, 4, 2, respectively.

We noted earlier that if two integers are congruent modulo m, then one of them is relatively prime to m if and only if the other is relatively prime to it. Consequently, every complete residue system modulo m contains exactly $\phi(m)$ elements relatively prime to m.

A set of elements in a complete residue system modulo m, which are relatively prime to m, is called a *reduced residue system* modulo m. Let S be a reduced residue system modulo m. Then, S contains exactly $\phi(m)$ elements; and an integer a is relatively prime to m if and only if a is congruent to some element in S. Moreover, if S is a set of $\phi(m)$ integers, any two elements of which are incongruent modulo m, then S is a reduced residue system modulo m.

Theorem 13 (Euler's extension of Fermat's theorem) *If n is a positive integer and a is an integer relatively prime to n, then $a^{\phi(n)} \equiv 1 \pmod{n}$.*

Proof Suppose that $S = \{b_1, \ldots, b_k\}$, where $k = \phi(n)$, is a reduced residue system modulo n, and a is an integer relatively prime to n. Then each of the products $ab_1, \ldots, ab_k$ is relatively prime to n. Moreover,

$$ab_i \equiv ab_j \pmod{n} \Longrightarrow b_i \equiv b_j \pmod{n},$$

by the cancellation law. Since $b_1, \ldots, b_k$ are incongruent mod n, this implies that $ab_1, \ldots, ab_k$ are k incongruent integers. Consequently, they are congruent modulo n to the elements in S, taken in some order. Multiplying these congruences we get,

$$(ab_1) \cdots (ab_k) \equiv b_1 \cdots b_k \pmod{n}.$$

Cancelling the product $b_1 \cdots b_k$, which is relatively prime to n, we get $a^k \equiv 1 \pmod{n}$. Since $k = \phi(n)$, the proof is complete. □

If p is prime, then $\phi(p) = p - 1$ and so Fermat's theorem is a special case of Euler's theorem.

A number-theoretic function $f(n)$ is said to be *multiplicative* if $f(mn) = f(m)f(n)$ for all m, n which are coprime.

Theorem 14 *The Euler function $\phi(n)$ is multiplicative, that is, $\phi(mn) = \phi(m)\phi(n)$ for all m, n with $\gcd(m, n) = 1$.*

Proof Since $\phi(1) = 1$, the result is trivial when $m = 1$ or $n = 1$. So let $m > 1$, $n > 1$ and $\gcd(m, n) = 1$. Write down the integers from 1 to mn in m rows and n columns as follows:

$$\begin{array}{ccccc} 1 & \ldots & j & \ldots & n \\ n+1 & \ldots & n+j & \ldots & 2n \\ \vdots & & \vdots & & \vdots \\ (m-1)n & \ldots & (m-1)n+j & \ldots & mn. \end{array}$$

The n entries in the ith row are $a_{i1}, a_{i2}, \ldots, a_{in}$, where

$$a_{ij} = (i-1)n + j, \quad 1 \leq i \leq m, \ 1 \leq j \leq n.$$

It is clear that for each i,

$$a_{ij} \equiv j \pmod{n},$$

and hence, $\gcd(a_{ij}, n) = 1$ if and only if $\gcd(j, n) = 1$. This shows that either all the entries a_{ij} in the jth column are relatively prime to n, or none of them is so. Moreover, there are $\phi(n)$ values of j in the range $1 \leq j \leq n$ which are relatively prime to n. Consequently, there are exactly $\phi(n)$ columns containing entries which are relatively prime to n.

If $a_{ij} \equiv a_{kj} \pmod{m}$, then

$$(i-1)n \equiv (k-1)n \pmod{m},$$

and so, $i-1 \equiv k-1 \pmod{m}$ as n is relatively prime to the modulus m. It follows that the m entries $a_{1j}, \ldots, a_{mj}$ of the jth column are incongruent modulo m, and hence they form a complete set of residues modulo m. It follows that each column contains $\phi(m)$ entries relatively prime to m. This holds for each of the $\phi(n)$ columns containing entries relatively prime to n. Hence, the total number of entries in the array which are relatively prime to

m and n, simultaneously, is $\phi(m)\phi(n)$. Since a number is relatively prime to mn if and only if it is relatively prime to m and n, the result follows. □

Another interesting property of the ϕ function is given below.

Theorem 15 *If n is any positive integer, then*

$$\sum_{d|n} \phi(d) = n,$$

where the summation is extended over all positive divisors of n.

Proof For a positive divisor d of n, let S_d denote the set of those integers k with

$$\gcd(k, n) = d, \quad 1 \le k \le n.$$

Let $\psi(d)$ be the number of elements in S_d. For each k with $1 \le k \le n$, there is one and only one positive integer d such that $\gcd(k, n) = d$. It follows that these sets S_d, corresponding to various divisors d of n, are pair-wise disjoint, and moreover, their union is the set $\{1, 2, \ldots, n\}$. Thus, we see that

$$\sum_{d|n} \psi(d) = n.$$

Now, for a divisor d of n, the integer k belongs to S_d only when $\gcd(k, n)$ is d, that is, $\gcd(k/d, n/d) = 1$. Thus, we see that k belongs to S_d if and only if $k = rd$ for some r relatively prime to n/d and $1 \le r \le n/d$; there are $\phi(n/d)$ such values of r. This shows that $\psi(d) = \phi(n/d)$ and hence,

$$\sum_{d|n} \phi(n/d) = n.$$

As d varies over the divisors of n, the numbers n/d also vary over the same divisors in another order. This proves the desired result. □

We shall next find a formula for finding the values of $\phi(n)$ for $n > 1$.

Theorem 16 *If p is prime and a is a positive integer, then*

$$\phi(p^a) = p^a - p^{a-1} = p^a\left(1 - \frac{1}{p}\right).$$

Proof Let $n = p^a$, and k be a positive integer not exceeding n. If k is not relatively prime to n, then $p|k$, and hence, $k = pm$ for some integer m with $1 \le m \le p^{a-1}$. This shows that there are exactly p^{a-1} positive integers not exceeding n and not relatively prime to n. Hence,

$$\phi(n) = n - p^{a-1} = p^a - p^{a-1}. \qquad \square$$

Corollary 16.1 *If $n > 1$, then $\phi(n) = n\prod_{p|n}(1 - 1/p)$, where the product is extended over all prime divisors p of n.*

Proof If $m_1, \ldots, m_k$ are pair-wise coprime, then m_{i+1} is relatively prime to the product $m_1 \cdots m_i$, and hence,

$$\phi(m_1 \cdots m_{i+1}) = \phi(m_1 \cdots m_i)\phi(m_{i+1}).$$

Using this for $1 \le i \le k - 1$, we get

$$\phi(m_1 \cdots m_k) = \phi(m_1) \cdots \phi(m_k).$$

Let $n = p_1^{a_1} \cdots p_k^{a_k}$, where $p_1, \ldots, p_k$ are the distinct prime divisors of n. The factors $p_1^{a_1}, \ldots, p_k^{a_k}$ of n are pair-wise coprime. So

$$\phi(n) = \phi(p_1^{a_1}) \cdots \phi(p_k^{a_k}) = \prod_{i=1}^{k} p_i^{a_1}\left(1 - \frac{1}{p_i}\right),$$

by the theorem. $\square$

For example,

$$\phi(100) = \phi(2^2 \cdot 5^2) = 100\left(1 - \frac{1}{2}\right)\left(1 - \frac{1}{5}\right) = 40.$$

1.4.2 Divisor functions

Definition The number of positive divisors of a positive integer n is denoted by $d(n)$. The sum of all these $d(n)$ divisors is denoted by $\sigma(n)$.

Note that 1 and n are always divisors of n, and that 1 is the only divisor of 1. Thus, $d(1) = 1$ and $\sigma(1) = 1$. Further, n is prime if and only if $d(n) = 2$, and $\sigma(n) = n + 1$.

Theorem 17 *Let $n = p_1^{n_1} p_2^{n_2} \cdots p_k^{n_k}$, where $p_1, \ldots, p_k$ are distinct primes and $n_1, \ldots, n_k$ are positive integers. Then,*

(i) $d(n) = (n_1+1)(n_2+1)\cdots(n_k+1)$,

(ii) $\sigma(n) = \dfrac{p_1^{n_1+1}-1}{p_1-1}\cdot\dfrac{p_2^{n_2+1}-1}{p_2-1}\cdots\dfrac{p_k^{n_k+1}-1}{p_k-1}$.

Proof By the fundamental theorem of arithmetic, every positive divisor a of n has a unique expression of the form $a = p_1^{a_1}\cdots p_k^{a_k}$, where a_i are integers with $0 \le a_i \le n_i$. Consider the product

$$\left(1+p_1+p_1^2+\cdots p_1^{n_1}\right)\left(1+p_2+p_2^2+\cdots p_2^{n_2}\right)\cdots$$
$$\left(1+p_k+p_k^2+\cdots p_k^{n_k}\right).$$

The various terms, obtained when the factors are multiplied out, are precisely the various divisors of n. Hence, the number of divisors of n is the number of terms in this continued product, and the sum of divisors is the value of this product. Since there are $(n_1+1)\cdots(n_k+1)$ terms in the product, we get (i). The value of each factor can be found using the formula for the sum of a geometric progression. This readily gives (ii). □

Theorem 18 *The divisor functions $d(n)$ and $\sigma(n)$ are multiplicative.*

Proof Let m and n be relatively prime positive integers. We have to prove that $d(mn) = d(m)d(n)$ and $\sigma(mn) = \sigma(m)\sigma(n)$. These are trivial when $m = 1$ or $n = 1$. So suppose that $m > 1$ and $n > 1$. Let

$$m = p_1^{m_1}\cdots p_k^{m_k}, \quad n = q_1^{n_1}\cdots q_r^{n_r}$$

be the standard representations of m and n as products of primes. Since m and n are relatively prime, we see that the primes p_i and q_j are all distinct. Hence, the standard representation of mn as a product of primes is

$$mn = p_1^{m_1}\cdots p_k^{m_k}\, q_1^{n_1}\cdots q_r^{n_r}.$$

Using the formula for the number of divisors we get

$$d(mn) = (m_1+1)\cdots(m_k+1)\,(n_1+1)\cdots(n_r+1).$$

The product of the first k factors is $d(m)$, and the product of the remaining r factors is $d(n)$. Hence, $d(mn) = d(m)d(n)$.
Similarly,

$$\begin{aligned}\sigma(mn) &= \frac{p_1^{m_1+1}-1}{p_1-1}\cdots\frac{p_k^{m_k+1}-1}{p_k-1}\,\frac{q_1^{n_1+1}-1}{q_1-1}\cdots\frac{q_r^{n_r+1}-1}{q_r-1}\\ &= \sigma(m)\sigma(n).\end{aligned}$$

□

An integer $n > 1$ is said to be *highly composite* if it has more divisors than any positive integer less than it, that is, if $d(m) < d(n)$ for $1 \leq m < n$. Note that the values of $d(n)$ for $n = 1, 2, 3, \ldots$ are $1, 2, 2, \ldots$. An examination of these values reveals that the first four highly composite numbers are 2, 4, 6 and 12.

1.4.3 The greatest integer function

For each real number x there is a unique integer m such that $m \leq x < m + 1$. This integer m is denoted by $[x]$.

Thus, $[1.3] = 1$, $[0.8] = 0$, $[-2.3] = -3$, and $x = [x] + t$, where $0 \leq t < 1$. Note that $[x] = x$ if and only if x is an integer. If $m \leq x < m + 1$ and $n \leq y < n + 1$, then $m + n \leq x + y < m + n + 2$. This shows that

$$[x + y] = [x] + [y] \quad \text{or} \quad [x] + [y] + 1.$$

Further, $[-x] = -[x] - 1$ if x is not an integer, and $[-x] = -[x]$ if x is an integer.

Theorem 19 *If x is any real number and n is a positive integer, then*

$$\left[\frac{x}{n}\right] = \left[\frac{[x]}{n}\right]$$

Proof Let $m = \left[\frac{x}{n}\right]$. Then,

$$m \leq \frac{x}{n} < m + 1, \quad mn \leq x < mn + n.$$

This implies that $[x] = mn + k$ for some integer k with $0 \leq k < n$. Thus,

$$mn \leq [x] = mn + k < mn + n,$$

and hence, $m \leq \frac{[x]}{n} < m + 1$. This shows that $\left[\frac{[x]}{n}\right] = m$. □

Theorem 20 *If n is a positive integer and p is a prime, then the highest power of p that divides $n!$ is p^e, where*

$$e = \left[\frac{n}{p}\right] + \left[\frac{n}{p^2}\right] + \left[\frac{n}{p^3}\right] + \cdots.$$

Proof Let $p^{k-1} \leq n < p^k$. We shall prove the result by induction on k. If $k = 1$, then $n < p$, and $n! = 1 \cdot 2 \cdot 3 \cdots n$ is relatively prime to p. So $e = 0$ in this case, and this agrees with the formula.

Let $k > 1$. In the product for $n!$, the factors which are divisible by p are p, $2p$, ..., mp, where m is an integer such that $mp \leq n < (m+1)p$, that is, $m = \left[\frac{n}{p}\right]$. If N is the product of the remaining factors, then N is relatively prime to p, and

$$n! = p^m \, m! \, N.$$

Hence, $e = m + a$, where p^a is the highest power of p that divides $m!N$. Since N is relatively prime to p, we see that p^a is the highest power of p that divides $m!$. Note that

$$mp \leq n < p^k \quad \text{and} \quad p^{k-1} \leq n < (m+1)p.$$

Hence, $m < p^{k-1}$ and $p^{k-2} < m + 1$. It follows that $p^{k-2} \leq m < p^{k-1}$. By induction hypothesis, we can use the formula to find a. Thus,

$$a = \left[\frac{m}{p}\right] + \left[\frac{m}{p^2}\right] + \cdots.$$

Since $\left[\frac{m}{p^i}\right] = \left[\frac{n}{p^{i+1}}\right]$, by the previous theorem, we get

$$e = m + a = m + \left[\frac{n}{p^2}\right] + \left[\frac{n}{p^3}\right] + \cdots.$$

This completes the proof. □

For example, if $n = 100$ and $p = 5$ then

$$e = \left[\frac{100}{5}\right] + \left[\frac{100}{5^2}\right] + \cdots = 20 + 4.$$

Thus, 5^{24} divides $100!$, and no higher power of 5 divides $100!$. Similarly, the highest power of 2 that divides $100!$ is $50 + 25 + 12 + 6 + 3 + 1 = 97$. It follows that $100!$ is divisible by 10^{24} but not by 10^{25}.

1.5 Quadratic residues

Let p be an odd prime, and a be an integer not divisible by p. If the quadratic congruence $x^2 \equiv a \pmod p$ has a solution, then a is said to be a *quadratic residue* of p. If this congruence has no solution, then a is said to be a *quadratic non-residue* of p.

In other words, a is a quadratic residue of p if and only if $r^2 \equiv a \pmod p$ for some r with $1 \le r \le p-1$. Since $(p-r)^2 \equiv r^2 \pmod p$, and r or $p-r$ is less than $p/2$, we see that a is a quadratic residue of p if and only if

$$r^2 \equiv a \pmod p \quad \text{for some } r \text{ with } 1 \le r \le (p-1)/2.$$

Suppose that $r^2 \equiv s^2 \pmod p$, where $1 \le r, s \le (p-1)/2$. Then,

$$(r-s)(r+s) = r^2 - s^2 \equiv 0 \pmod p.$$

Since $|r-s| \le (p-1)/2$ and $2 \le r+s \le p-1$, this holds only when $r = s$. It follows that the quadratic residues are precisely those which are congruent modulo p to

$$1^2, 2^2, \ldots, \left(\frac{p-1}{2}\right)^2.$$

A reduced residue system modulo p has $\phi(p) = p-1$ elements. It contains exactly $(p-1)/2$ quadratic residues; the remaining $(p-1)/2$ elements are non-residues.

For example, for $p = 5$, the quadratic residues are 1 and 4; and the non-residues are 2 and 3. For $p = 11$, the residues are 1, 4, 9, $16 \equiv 5$ and $25 \equiv 3$; and the non-residues are 2, 6, 7, 8 and 10.

Theorem 21 **(Euler's criterion)** *Let p be an odd prime and a be an integer relatively prime to p. Then a is a quadratic residue or non-residue of p according as $a^{(p-1)/2}$ is congruent modulo p to 1 or -1.*

Proof Suppose that a is a residue of p. Then, $a \equiv r^2 \pmod p$ for some r with $1 \le r \le (p-1)/2$, and

$$a^{(p-1)/2} - 1 \equiv r^{p-1} - 1 \equiv 0 \pmod p$$

by Fermat's theorem. Since there are $(p-1)/2$ incongruent residues, this shows that the congruence

$$x^{(p-1)/2} - 1 \equiv 0 \pmod{p} \tag{*}$$

has at least $(p-1)/2$ solutions. However, this polynomial congruence cannot have more that $(p-1)/2$ incongruent solutions by Lagrange's theorem [Theorem **??**]. It follows that the quadratic residues of p are precisely the roots of (*).

Next suppose that a is a non-residue of p. Then, a does not satisfy (*). However,

$$\left(a^{(p-1)/2} - 1\right)\left(a^{(p-1)/2} + 1\right) = a^{p-1} - 1$$

is divisible by p using Fermat's theorem. Since the first factor cannot be divisible by p, the other factor must be divisible by p. That is, $a^{(p-1)/2} \equiv -1 \pmod{p}$.

Theorem 22 **(Gauss' lemma)** *Let p be an odd prime, and a be an integer not divisible by p. Let $r_1, r_2, \ldots, r_{(p-1)/2}$ be integers such that*

$$ia \equiv r_i \pmod{p}, \quad |r_i| < \frac{p}{2}, \quad 1 \le i \le \frac{p-1}{2}.$$

If μ of these integers r_i are negative, then

$$a^{(p-1)/2} \equiv (-1)^{\mu} \pmod{p}.$$

Proof Write $k = (p-1)/2$. Since a is not divisible by p, we see that its multiples ia, $1 \le i \le k$ are incongruent modulo p, and none of them is divisible by p. Hence, $0 < |r_i| < p/2$. Further,

$$r_1 r_2 \cdots r_k \equiv a(2a) \cdots (ka) \pmod{p}. \tag{*}$$

Suppose that $|r_i| = |r_j|$ for some i, j. Then, $r_i^2 = r_j^2$, and hence,

$$(i-j)(i+j)a^2 = i^2a^2 - j^2a^2 \equiv 0 \pmod{p}.$$

Since $|i-j| < p/2$, $2 \le i+j \le 2k = p-1$ and $\gcd(a,p) = 1$, this implies that $i = j$. It follows that $|r_1|, \ldots, |r_k|$ are $k = (p-1)/2$ distinct integers between 0 and $p/2$. So their product is $k\,!$. But their product is also equal to $(-1)^{\mu} r_1 \cdots r_k$. Hence,

$$k\,! = (-1)^{\mu} r_1 \cdots r_k \equiv (-1)^{\mu} a^k k\,! \pmod{p}.$$

by (*). Cancelling $k!$ which is relatively prime to p, we get

$$1 \equiv (-1)^{\mu} a^k \pmod{p}, \quad a^k \equiv (-1)^{\mu} \pmod{p},$$

that is, $a^{(p-1)/2} \equiv (-1)^{\mu} \pmod{p}$. □

As a simple application of Gauss' lemma let us examine whether 7 is a quadratic residue or non-residue of $p = 19$. Consider the first $(p-1)/2$ multiples of 7. They are $7, 14, \ldots, 63$. Their residues with the least absolute values are

$$7, -5, 2, 9, -3, 4, -8, -1, 6,$$

and 4 of them are negative. So $\mu = 4$, and $7^9 \equiv 1 \pmod{19}$. By Euler's criterion [Theorem ??], 7 is a quadratic residue of 19. It may be noted that $7 \equiv 8^2 \pmod{19}$.

Corollary 22.1 *For an odd prime p we have*

$$2^{(p-1)/2} \equiv (-1)^{[(p+1)/4]} \pmod{p}.$$

Proof Let $m = [(p+1)/4]$. Then,

$$m \leq \frac{p+1}{4} < m+1, \quad 4m \leq p+1 < 4m+4.$$

Since p is odd, we must have $p+1 = 4m$ or $4m+2$.

Use Gauss' lemma. We have to consider the residues of multiples of 2 having the least absolute value. Let $r_1, r_2, \ldots, r_{(p-1)/2}$ be given by

$$2i \equiv r_i \pmod{p}, \quad |r_i| < \frac{p}{2}, \quad 1 \leq i \leq \frac{p-1}{2}.$$

Clearly, $0 < 2i < p$ for $1 \leq i \leq (p-1)/2$ and $r_i = 2i$ or $2i - p$. Precisely,

$$r_i = \begin{cases} 2i & \text{for } 0 < 2i < p/2 \\ 2i - p & \text{for } p/2 < 2i < p. \end{cases}$$

So $r_i < 0$ if and only if $p/4 < i < p/2$. If $p+1 = 4m$, then this holds if and only if $m \leq i \leq 2m-1$. If $p+1 = 4m+2$, then this holds if and only if $m+1 \leq i \leq 2m$. In either case, there are exactly m values of i for which $r_i < 0$. Hence,

$$2^{(p-1)/2} \equiv (-1)^m \pmod{p}$$

by Gauss' lemma. □

Corollary 22.2 *Let p be a prime. If $p \equiv \pm 1 \pmod 8$, then 2 is a quadratic residue of p, and if $p \equiv \pm 3 \pmod 8$, then 2 is a non-residue of p.*

Proof If $p = 8k \pm 1$, then $[(p+1)/4] = 2k$, and if $p = 8k \pm 3$, then $[(p+1)/4] = 2k - 1$ or $2k + 1$. Hence, $(-1)^{[(p+1)/4]}$ is equal to 1 in the former case, and to -1 in the other case. Using the previous corollary, we see that $2^{(p-1)/2}$ is congruent modulo p to 1 if $p = 8k \pm 1$ and to -1 if $p = 8k \pm 3$. An application of Euler's criterion [Theorem ??] yields the result. □

1.6 Primitive Roots

If m is a positive integer and $\gcd(a, m) = 1$, then $a^{\phi(m)} \equiv 1 \pmod m$. Thus, there are positive integers k such that $a^k \equiv 1 \pmod m$, $\phi(m)$ being one of them. The smallest positive integer h with the property that $a^h \equiv 1 \pmod m$ is called the *order* or *exponent* of a modulo m. Thus, $1 \leq h \leq \phi(m)$. Moreover, if $\gcd(a, m) = 1$ and $a \equiv b \pmod m$, then a and b have the same order modulo m.

Definition Let m be a positive integer. A *primitive root* of m is a positive integer a such that $\gcd(a, m) = 1$ and the order of a modulo m is $\phi(m)$; that is, there is no integer k with $1 \leq k < \phi(m)$ and $a^k \equiv 1 \pmod m$.

Theorem 23 *If $\gcd(a, m) = 1$ and a has order h modulo m, then*

(a) $a^n \equiv 1 \pmod m$ *if and only if* $n \equiv 0 \pmod h$,

(b) $a^i \equiv a^j \pmod m \Longrightarrow j - i \equiv 0 \pmod h$.

Proof (a) If $n = kh$, then $a^n = (a^h)^k \equiv 1 \pmod m$. On the other hand, suppose that $a^n \equiv 1 \pmod m$. Let $n = kh + r$, where $0 \leq r < h$. Since

$$a^n = (a^h)^k \cdot a^r \equiv a^r \pmod m,$$

we see that $a^r \equiv 1 \pmod m$. This shows that r cannot be a positive integer, as h is the smallest such positive integer. Hence $r = 0$, and $n = kh$.

(b) If $a^i \equiv a^j \pmod m$, where $i \leq j$, then

$$a^j = a^i \cdot a^{j-i} \equiv a^j \cdot a^{j-i} \pmod m,$$

which implies that $a^{j-i} \equiv 1 \pmod{m}$, and so $j - i \equiv 0 \pmod{h}$ by (a). □

Corollary 23.1 *If the integer a has order h modulo m, then $a, a^2, \ldots, a^h$ are incongruent modulo m.*

Proof If $a^i \equiv a^j \pmod{m}$, where $1 \leq i \leq j \leq h$, then $j - i \equiv 0 \pmod{h}$, $0 \leq j - i < h$, and hence, $j = i$. This proves the result. □

Theorem 24 *Let a be relatively prime to m and the order of a modulo m be h. If k is a positive integer and $\gcd(k, h) = d$, then the order of a^k modulo m is h/d. Consequently, the order of a^k modulo m is h if and only if $\gcd(k, h) = 1$.*

Proof Let t be the order of a^k modulo m. Since

$$\left(a^k\right)^{h/d} = \left(a^h\right)^{k/d} \equiv 1 \pmod{m},$$

this implies that h/d is divisible by t. On the other hand, $a^{kt} \equiv 1 \pmod{m}$, and hence,

$$kt \equiv 0 \pmod{h}, \quad \frac{k}{d}t \equiv 0 \pmod{\frac{h}{d}}.$$

Since h/d is relatively prime to k/d, this implies that t is divisible by h/d. It follows that $t = h/d$. This completes the proof. □

Theorem 25 *If p is a prime and d is a positive divisor of $p - 1$, then there are $\phi(d)$ incongruent integers having order d modulo p.*

Proof Let $S = \{1, 2, \ldots, p-1\}$. For a positive divisor d of $p-1$, let S_d denote the set of those elements in S having order d modulo p. Let $\psi(d)$ be the number of elements in S_d, where we take $\psi(d) = 0$ if S_d contains no element. We have to prove that $\psi(d) = \phi(d)$ for all d.

Every element a in S has a unique order d, and since $a^{p-1} \equiv 1 \pmod{p}$, we have $d|p-1$ [Theorem ??(a)]. It follows that the sets S_d corresponding to distinct divisors d are disjoint, and their union is S. Consequently,

$$\sum_{d|p-1} \psi(d) = p - 1,$$

where the summation is extended over the positive divisors d of $p-1$. Since $\sum_{d|p-1}\phi(d)=p-1$ by Theorem **??**, this shows that

$$\sum_{d|p-1}\{\phi(d)-\psi(d)\}=0. \tag{1}$$

Consider a divisor d of $p-1$ for which the set S_d is non-empty. Let a be an element of S_d. Then, a has order d, and

$$(a^k)^d=(a^d)^k\equiv 1 \pmod p.$$

Thus, $a, a^2, \ldots, a^d$ satisfy the congruence

$$x^d-1\equiv 0 \pmod p. \tag{2}$$

By Lagrange's theorem, this cannot have more than d incongruent roots. Since $a, a^2, \ldots, a^d$ are incongruent modulo p, by the corollary of Theorem **??**, we see that (2) has precisely d incongruent roots $a, a^2, \ldots, a^d$.

Let b be any element of S_d. Then, the order of b modulo p is d, and hence b satisfies (2). This implies that b is congruent modulo p to one of the numbers a^k with $1\le k\le d$. Hence, b and a^k have the same order modulo p. But the order of a^k is equal to d/d_1, where $d_1=\gcd(k,d)$, by Theorem **??**. This shows that $d=d/d_1$, and hence, $\gcd(k,d)=d_1=1$. Thus, we have proved that corresponding to each b in S_d, there is an integer k such that

$$b\equiv a^k \pmod p, \quad \text{where } 1\le k\le d,\ \gcd(k,d)=1.$$

Distinct elements b in S_d will correspond to distinct values of k since elements of S_d are incongruent modulo p. Since the number of values of k satisfying

$$\gcd(k,d)=1, \quad 1\le k\le d,$$

is $\phi(d)$, we see that S_d cannot contain more than $\phi(d)$ elements. Thus, $\psi(d)\le\phi(d)$ for all divisors d for which S_d is non-empty. This being trivial when S_d is empty, we see that each term in (1) is non-negative. This implies that $\phi(d)-\psi(d)=0$ for all divisors d of $p-1$, as we desired. □

Corollary 25.1 *Every prime p has $\phi(p-1)$ incongruent primitive roots.*

1.7 Miscellaneous

As stated earlier, polynomial equations in two or more unknowns, whose solutions are required to be integers, are known as *Diophantine equations*. Linear equations in two unknowns and some higher degree equations have been studied in detail.

Second degree equations of the type $x^2 - dy^2 = \pm 1$, where d is a positive integer which is not a perfect square, have been studied extensively. These are known as *Pell's equations*.

Consider the equation $x^2 - dy^2 = N$. If $d \leq 0$, then this can have at the most a finite number of solutions. If $d = a^2$, a perfect square, then this becomes $(x - ay)(x + ay) = N$, which can also have only a finite number of solutions. Suppose that d is a positive integer that is not a perfect square. It has been proved that the equation $x^2 - dy^2 = \pm 1$ has always a solution in integers. If $x_1^2 - dy_1^2 = \pm 1$, then we can find any number of solutions of $x^2 - dy^2 = \pm 1$ as follows: Define x_n, y_n successively by

$$x_{n+1} = x_1 x_n + dy_1 y_n, \quad y_{n+1} = x_1 y_n + x_n y_1 \quad \text{for } n \geq 1.$$

Then,

$$x_{n+1}^2 - dy_{n+1}^2 = (x_1^2 - dy_1^2)(x_n^2 - dy_n^2).$$

Thus, $x_n^2 - dy_n^2 = \left(x_1^2 - dy_1^2\right)^n$.

1.7.1 Pythagorean triples

If x, y, z are positive integers such that $x^2 + y^2 = z^2$, then (x, y, z) is said to be a *Pythagorean triple*. It is said to be *primitive* if there is no common divisor greater than 1 for x, y, z.

Let x, y, z be integers, not all zero, such that $x^2 + y^2 = z^2$. If d divides two of them, then d^2 divides the square of the third. Note that if $(a/b)^2$ is an integer, then a/b is also an integer. Hence, $b^2|a^2$ implies that $b|a$. Thus, if d divides two of x, y, z, then it also divides the third. It follows that if the positive integers x_1, y_1, z_1 form a primitive Pythagorean triple, then they are pair-wise relatively prime. Moreover, given any Pythagorean triple (x, y, z) we can find a primitive triple (x_1, y_1, z_1) and a positive integer k such that $x = kx_1$, $y = ky_1$ and $z = kz_1$.

Theorem 26 *Primitive Pythagorean triples (x, y, z) are of the form*

$$x = r^2 - s^2, \quad y = 2rs, \quad z = r^2 + s^2,$$

where r, s are positive integers, one of which is odd and the other even, $r > s$, and $\gcd(r, s) = 1$.

Proof Let (x, y, z) be a primitive Pythagorean triple. Then, x, y, z are pair-wise coprime. The square of an integer is of the form $4k$ or $4k + 1$ according as the integer is even or odd. If x and y are odd, then $x^2 + y^2$ is congruent to 2 modulo 4, and so cannot be the square of an integer. Hence, x or y is even. Suppose that y is even. Then, x and z are odd since x, y, z are pair-wise coprime. Thus, $z + x$ and $z - x$ are even. Let $z + x = 2a$ and $z - x = 2b$. Then,

$$a + b = z, \quad a - b = x, \quad y^2 = z^2 - x^2 = 4ab.$$

Let $d = \gcd(a, b)$. Then, d divides z and x, and so $d = 1$. Since $ab = (y/2)^2$ is the square of an integer, this implies that a and b are perfect squares. Let $a = r^2$ and $b = s^2$, where r and s are positive integers. Then, $\gcd(r, s) = 1$ as $\gcd(a, b) = 1$, and

$$x = a - b = r^2 - s^2, \quad z = a + b = r^2 + s^2,$$

$$y^2 = 4ab = (2rs)^2.$$

If r and s are odd, then x will be even. Hence, $r \not\equiv s \pmod 2$, and further, $r > s$.

Finally, if x, y, z are given like this, then

$$x^2 + y^2 = (r^2 - s^2)^2 + 4r^2s^2 = (r^2 + s^2)^2 = z^2.$$

Let $d = \gcd(x, z)$, then d divides $z + x = 2r^2$ and $z - x = 2s^2$. Since $r \not\equiv s \pmod 2$, we see that $x = (r + s)(r - s)$ is odd, and hence, d is odd as $d | x$. It follows that d divides r^2 and s^2. Hence, $d = 1$ as $\gcd(r^2, s^2) = 1$. This shows that (x, y, z) is primitive, and completes the proof. □

We cannot so easily deal with similar Diophantine equations of the form $x^n + y^n = z^n$, where $n > 2$. Fermat stated that if $n > 2$, then this has no solution in integers other than the trivial ones in which x or y is zero. This statement, known as *Fermat's last theorem*, has been proved to be true.

However, there are similar equations having solutions in integers. Let a, b, c be a positive integers and c be relatively prime to a and b. Then there is a positive integer k such that $kab \equiv -1 \pmod{c}$. If $kab + 1 = Nc$, let

$$x = 2^{(k+rc)b}, \quad y = 2^{(k+rc)a}, \quad z = 2^{N+rab},$$

where r is some positive integer. Then,

$$x^a = 2^{(k+rc)ab}, \quad y^b = 2^{(k+rc)ab},$$

$$z^c = 2^{Nc+rabc} = 2^{kab+1+rabc}.$$

We have $2^m + 2^m = 2^{m+1}$. Using this with $m = (k + rc)ab$, we get

$$x^a + y^b = z^c.$$

Since r can be any positive integer, we get infinitely many solutions of this. Thus, $x^n + y^n = z^{n+1}$ has infinitely many non-trivial solutions.

The nth *Fibonacci number* u_n is defined by

$$u_1 = 1, \quad u_2 = 1, \quad u_{n+2} = u_{n+1} + u_n \quad \text{for } n \geq 1.$$

The first seven terms in the Fibonacci sequence $\{u_n\}$ are 1, 1, 2, 3, 5, 8 and 13.

The nth *triangular number* t_n is the sum of the first n natural numbers. Thus,

$$t_1 = 1, \quad t_2 = 3, \quad t_n = n(n+1)/2 \quad \text{for } n \geq 1.$$

A positive integer n is said to be *perfect* if the sum of its positive divisors, including 1 but excluding n, is equal to n, that is, if $\sigma(n) = 2n$. The positive divisors of 6, other than 6, are 1, 2 and 3 whose sum is 6. The positive divisors of 28, other than 28, are 1, 2, 4, 7, and 14 whose sum is 28. Thus, 6 and 28 are perfect.

Mersenne numbers

The numbers $M_n = 2^n - 1$ are called *Mersenne numbers*. Thus, if $n = 2$, 3, 5 and 7, then $M_n = 3, 7, 31$ and 127 respectively. These are primes. But $M_{11} = 2047 = 23 \times 89$ is composite.

If a and n are integers greater than 1, then

The second factor is obviously greater than 1. So if $a^n - 1$ is a prime, then the first factor is necessarily 1, that is, $a = 2$. Thus, primes of the form $a^n - 1$ are necessarily Mersenne primes.

Moreover, suppose that M_n is prime, and $n = ab$, where a and b are positive integers. Then, $M_n = 2^{ab} - 1$ is divisible by $2^a - 1$ and $2^b - 1$. If $a > 1$ and $b > 1$, then these factors $2^a - 1$ and $2^b - 1$ are greater than 1. Hence, if M_n is prime, then a or b must be 1, that is, n must be prime.

Fermat numbers

Suppose that m is a positive integer having an odd prime divisor p. If $m = kp$, then $2^m + 1 = (2^k)^p + 1$ is divisible by $2^k + 1$. Since $1 < 2^k + 1 < 2^m + 1$, we see that $2^m + 1$ cannot be prime in this case. Hence, if $2^m + 1$ is prime, then m is a power of 2.

The nth *Fermat number* F_n is defined by

$$F_n = 2^{2^n} + 1 \quad \text{for } n \geq 0.$$

The first five Fermat numbers are

$$F_0 = 3, \quad F_1 = 5, \quad F_2 = 17, \quad F_4 = 257.$$

These are primes. However, F_5 is not prime; it is divisible by the prime $p = 641$. This can be seen as follows: We have

$$5 \times 2^7 = 640 \equiv -1 \pmod{p}, \quad 5^4 \times 2^{28} \equiv 1 \pmod{p}.$$

Since $5^4 \equiv -16 \pmod{p}$, we get

$$(-16)2^{28} \equiv 1 \pmod{p}, \quad 2^{32} + 1 \equiv 0 \pmod{p},$$

that is, $F_5 \equiv 0 \pmod{641}$.

TWO

Problems

Set I

1. The nth *triangular number* t_n is the sum of the first n natural numbers. Thus, $t_n = n(n+1)/2$ for $n = 1, 2, \ldots$. If the nth triangular number t_n is a perfect square, prove that $t_{4n(n+1)}$ is also a perfect square. Deduce that there are an infinite number of triangular numbers which are perfect squares.

2. Find three consecutive triangular numbers t_n, t_{n+1}, t_{n+2}, whose product is a perfect square. Also, show that there exists an infinite number of triples of consecutive triangular numbers whose product is a perfect square.

3. Verify that 10 is the only triangular number that can be written as the sum of two consecutive odd squares.

4. Let n be a positive integer. If $a_1, \ldots, a_n$ are integers and the equation

$$x^n + a_1x^{n-1} + a_2x^{n-2} + \cdots + a_n = 0$$

has a rational root, prove that the root must be an integer.

5. If $(a, 4) = 2 = (b, 4)$, prove that $(a + b, 4) = 4$.

6. If a, b, c are positive integers such that $(a, b, c)[a, b, c] = abc$, prove that $(a, b) = (b, c) = (a, c) = 1$.

7. Prove that $[a, b, c](ab, bc, ca) = abc$ for all positive integers a, b, c.

8. If m, n are positive integers such that $3m + n = 3[m, n] + (m, n)$, prove that n divides m.

9. Examine whether the following statements are true or false. If true, prove the result, and if false, give a counterexample:

   (a) If $(a, b) = 1$, then $(a^2, ab, b^2) = 1$.
   (b) If $(a, b) = 1$, then $(a + b, a^2 - ab + b^2)$ is either 1 or 3.
   (c) If $b|(a^2 + 1)$, then $b|a^4 + 1$.
   (d) Every positive integer can be written in the form $p + a^2$, where p is a prime or 1, and $a \geq 0$.
   (e) If p is prime and $p > 3$, then $p^2 + 2$ is composite.

10. If $\{p_n\}$ is a sequence of distinct primes, prove that

$$\frac{1}{p_1} + \frac{1}{p_2} + \cdots + \frac{1}{p_n}$$

is never an integer.

11. Let $R_n = 11 \cdots 1$ in decimal notation, where the digit 1 is repeated n times. Prove that

   (a) $R_n | R_m \iff n | m$,
   (b) if $(n, m) = 1$, then $(R_n, R_m) = 1$.
   (c) Find all n such that R_n is a perfect square.

12. Find the least integer $n > 1$ such that the number $9997n$ contains only odd digits in decimal notation.

13. Find the least positive integer n such that the product of 9999 and n contains only the digit 1 in decimal notation.

14. Prove that the sequence

$$(n + 1)! - 2, (n + 1)! - 3, \ldots, (n + 1)! - (n + 1)$$

produces n consecutive composite numbers for $n > 2$. Prove also that each term in this sequence is divisible by a prime which does not divide any other term in it.

15. Prove that for any positive integers n and k, there exists a set of n consecutive integers such that each member of this set is divisible by k distinct primes, no one of which divides any other member of the set.

16. Consider an arithmetic progression $p, p+d, \ldots, p+(n-1)d$. Prove that if all the n terms are prime numbers, and $n > 2$, then the common difference d is divisible by every prime $q < n$.

17. Find an arithmetic progression of 6 terms, each of which is a prime.

18. Let k be a positive integer. Prove that the arithmetic progression

$$a, a+d, a+2d, a+3d, \ldots$$

of positive integers, where $(a, d) = 1$, contains k consecutive terms that are composite.

19. Given any positive integer k, prove that there are k consecutive integers each divisible by a square greater than 1.

20. Use the Chinese remainder theorem to find five consecutive terms in the arithmetic progression $1, 6, 11, 16, \ldots$, each of which is divisible by a prime which does not divide another one.

21. Determine all primes that can divide two successive integers of the form $n^2 + 3$.

22. The nth triangular number t_n is the sum of the first n natural numbers. Determine all twin primes $p, p+2$ whose arithmetic mean is a triangular number.

23. Determine all twin primes p and $q = p + 2$ such that $pq - 2$ is also prime.

24. (a) Using Bertrand's theorem, or otherwise, show that if $n > 1$, then $n\,!$ is never a perfect square.
    (b) Find all positive integers n such that $n\,! + (n+1)\,! + (n+2)\,!$ is a perfect square.

25. For $n \equiv \pm 4 \pmod 9$, show that the equation $n = a^3 + b^3 + c^3$ has no integer solutions.

26. Show that each term of the sequence

$$16, 1156, 111556, 11115556, 1111155556, \ldots$$

is a perfect square.

27. Find all primes of the form $p^2 + 2^p$, where p is prime.

28. The primes $37, 67, 73, 79, \ldots$ are of the form $p = 36ab + 6a - 6b + 1$, with $a \geq 1$, $b \geq 1$. Show that no pair of twin primes can contain a prime of this form.

29. Let $p_1, p_2, \ldots, p_n$ be the first n primes in the natural order. Show, for $n \geq 2$, that there are at least two new primes in the interval $p_n < x \leq p_1 p_2 \cdots p_n + 1$.

30. (a) Find the last two digits in the decimal representation of the number 9^{9^9}.

    (b) Find the last three digits in the decimal representation of the number $9^{9^{9^9}}$.

Set II

1. Given integers x and y, let $m = ax + by$ and $n = cx + dy$, where a, b, c, d are integers such that $ad - bc = \pm 1$. Prove that $\gcd(m, n) = \gcd(x, y)$.

2. If a, m, n are positive integers and $m \neq n$, find $\gcd(a^{2^m} + 1, a^{2^n} + 1)$.

3. Prove that every integer $n \geq 12$ is the sum of two composite numbers.

4. Let a, b, x, y be positive integers such that $\gcd(a, b) = 1$ and $x^a = y^b$. Prove that $x = n^b$ and $y = n^a$ for some n.

5. Let a, b, n be positive integers such that $\gcd(a, b) = 1$ and $n > ab$. Prove that there exist positive integers x and y such that $n = ax + by$. Prove also that this is not true if $n = ab$.

6. If $n > 1$ is an integer not of the form $6k + 3$, prove that $n^2 + 2^n$ is composite.

7. If p is any prime other than 2 or 5, prove that

   (a) p divides infinitely many terms in the sequence $9, 99, 999, \ldots$.

   (b) p divides infinitely many terms in the sequence $1, 11, 111, \ldots$.

8. For any prime p, if $a^p \equiv b^p \pmod{p}$, prove that $a^p \equiv b^p \pmod{p^2}$.

9. Given an integer n, prove that there is an integer m divisible by n such that the decimal representation of m contains only the digits 0 and 1. Prove also that the same is true for digits 0 and 2, or 0 and 3, . . ., or 0 and 9, but for no other pair of digits.

10. For $n \geq 1$, show that

$$(-13)^{n+1} \equiv (-13)^n + (-13)^{n-1} \pmod{181}.$$

Determine all integers a such that the congruence $a^{n+1} \equiv a^n + a^{n-1} \pmod{181}$ holds for all positive integers n.

11. Prove that if a is an odd integer, then $a^{2^n} \equiv 1 \pmod{2^{n+2}}$ for all positive integers n.

12. If p and q are distinct primes, prove that $p^{q-1} + q^{p-1} \equiv 1 \pmod{pq}$.

13. Show that there exist no primes p and q that satisfy the condition $p^2 = 10^q - 999$.

14. Find the last two digits in the decimal representation of $3^{1001} + 2^{1001}$.

15. Let p be an odd prime, and $r_1, r_2, \ldots, r_{p-1}$ be the integers $1, 2, \ldots, p-1$ arranged in some order. Prove that at least two of the numbers

$$r_1,\ 2r_2,\ \ldots,\ (p-1)r_{p-1}$$

are congruent modulo p.

16. If n is a positive integer and x is real, prove that

$$[x] + \left[x + \frac{1}{n}\right] + \cdots + \left[x + \frac{n-1}{n}\right] = [nx].$$

17. Determine the set of all real numbers α such that

$$\left[\sqrt{n+\alpha} + \tfrac{1}{2}\right] = \left[\sqrt{n} + \tfrac{1}{2}\right] \quad \text{for all positive integers } n.$$

18. If $\gcd(m, n) = 1$, prove that

$$\sum_{k=1}^{n-1} \left[\frac{mk}{n}\right] = \frac{(m-1)(n-1)}{2}.$$

19. Let θ be a real number such that $0 < \theta < 1$. Define

$$x_n = \begin{cases} 0 & \text{if } [n\theta] = [(n-1)\theta], \\ 1 & \text{otherwise.} \end{cases}$$

Prove that

$$\lim_{n\to\infty} \frac{x_1 + x_2 + \cdots + x_n}{n} = \theta.$$

20. Let n be an odd positive integer. If n factors into the product of two positive integers, $n = uv$, with $u > v$ and $u - v \le \sqrt[4]{64n}$, prove that the roots of the equation $x^2 - 2\left[\sqrt{n} + 1\right]x + n = 0$ are integers.

21. Let α and β be positive irrational numbers such that $\dfrac{1}{\alpha} + \dfrac{1}{\beta} = 1$. Prove that the two infinite sequences

$$[\alpha], [2\alpha], [3\alpha], \ldots \quad \text{and} \quad [\beta], [2\beta], [3\beta], \ldots$$

together contain each natural number exactly once.

22. Let α and β be positive real numbers such that the two infinite sequences

$$[\alpha], [2\alpha], [3\alpha], \ldots \quad \text{and} \quad [\beta], [2\beta], [3\beta], \ldots$$

together contain each natural number exactly once. Prove that α and β are irrational numbers such that $\dfrac{1}{\alpha} + \dfrac{1}{\beta} = 1$.

23. Let p be prime, k a positive integer, and $1 \le n \le p^k - 1$. Prove that

(a) $\dbinom{p^k}{n} \equiv 0 \pmod{p}$.

(b) If, further, n is not divisible by p^{k-1}, then $\dbinom{p^k}{n} \equiv 0 \pmod{p^2}$.

24. (a) Determine all positive integers n such that $\left[\sqrt{n}\right]$ divides n.

(b) Determine all positive integers n, not a perfect square, such that $\left[\sqrt{n}\right]^3$ divides n^2.

25. Prove that $[3x] + [3y] \ge [x] + [y] + 2[x + y]$ for all real numbers x, y.

26. Prove that $[4x] + [4y] \ge [x] + [y] + [2x + y] + [2y + x]$ for real numbers x, y.

27. Prove that $[5x] + [5y] \geq [x] + [y] + [3x + y] + [3y + x]$ for all real numbers x, y.

28. Determine all positive integers k with the property that

$$[kx] + [ky] \geq [x] + [y] + [(k-2)x + y] + [(k-2)y + x]$$

for all real numbers x and y.

29. Let m and n be positive integers. Prove that

$$\frac{(3m)!\,(3n)!}{m!\,n!\,(m+n)!\,(m+n)!}, \quad \frac{(4m)!\,(4n)!}{m!\,n!\,(2m+n)!\,(2n+m)!},$$
$$\frac{(5m)!\,(5n)!}{m!\,n!\,(3m+n)!\,(3n+m)!}$$

are integers, and that

$$\frac{(6m)!\,(6n)!}{m!\,n!\,(4m+n)!\,(4n+m)!}$$

is not necessarily an integer.

30. Determine all integers $n > 1$ such that $(n-2)!$ is not divisible by n^2.

Set III

1. Prove that $\frac{1}{2}\sqrt{n} \leq \phi(n) \leq n$ for all n.

2. Prove that for a fixed integer n, the equation $\phi(x) = n$ has only a finite number of solutions.

3. Prove that

   (a) if n has r distinct prime divisors, then $\phi(n) \geq n/2^r$,

   (b) if n has r distinct odd prime divisors, then $\phi(n)$ is divisible by 2^r.

4. Prove that there is no solution for the equation $\phi(x) = 14$, and that 14 is the least even integer with this property. Apart from 14, what is the next smallest even integer n such that $\phi(x) = n$ has no solution?

5. Find all positive integers n for which $\phi(n) \equiv 2 \pmod 4$.

6. Prove that, for $n \geq 2$, the sum of all positive integers less than n and relatively prime to n is $\frac{1}{2}n\phi(n)$.

7. Define $f(n)$ as the sum of the positive integers less than n and prime to n. Prove that $f(m) = f(n)$ implies that $m = n$.

8. Determine all positive integers n such that

   (a) $\phi(n) = n/2$,

   (b) $\phi(n) = n/3$.

9. Prove that if k is an integer greater than 3, then there are no integers n for which $\phi(n) = n/k$.

10. Find all integers n for which $2^n - 1$ is divisible by 7. Also, show that there is no positive integer n for which $2^n + 1$ is divisible by 7.

11. Let n be a positive integer which is not a square. Prove that for every integer a relatively prime to n, there exist integers x and y satisfying

$$ax \equiv y \pmod n \quad \text{with } 0 < |x| < \sqrt{n} \text{ and } 0 < |y| < \sqrt{n}.$$

12. Let p be an odd prime. Prove the following:

   (a) If $q = (p-1)/2$, then $(q\,!)^2 + (-1)^q \equiv 0 \pmod p$.

   (b) $1^2 3^2 5^2 \cdots (p-2)^2 \equiv (-1)^{(p+1)/2} \pmod p$.

   (c) $2^2 4^2 6^2 \cdots (p-1)^2 \equiv (-1)^{(p+1)/2} \pmod p$.

13. Let p be a prime and $p \equiv 1 \pmod 4$. Prove that there exist positive integers x and y such that $p = x^2 + y^2$. Show also that such an expression of p is unique except for the order of summands.

14. Consider a prime $p \equiv 1 \pmod{60}$. Show that there exist positive integers a and b with $p = a^2 + b^2$, where 3 divides a or b and 5 divides a or b.

15. Let the positive integer n be written as $n = N^2 m$, where m is square-free. Prove that n can be represented as the sum of two squares if and only if m has no prime divisor of the form $4k + 3$.

16. Let $n = a^2 + b^2$, where a and b are positive integers with $\gcd(a, b) = 1$. Prove that any positive divisor of n is a sum of two squares. Deduce that there are infinitely many primes of the form $4k + 1$.

17. Given an odd integer $n \geq 3$, let k and t be the least positive integers such that both $kn + 1$ and tn are perfect squares. Prove that n is prime if and only if both k and t are greater than $n/4$.

18. Let s_n denote the sum of the first n primes. Prove that, for each n, there is an integer whose square lies between s_n and s_{n+1}.

19. Find all positive integers n for which $(n-1)! + 1$ is a power of n.

20. Prove that if n is a positive integer such that the equation

$$x^3 - 3xy^2 + y^3 = n$$

has a solution in integers (x, y), then it has at least three such solutions. Show also that if n is congruent modulo 9 to 2, 4, 5 or 7, then the equation has no solution in integers.

21. Let p be a prime, $p \geq 5$ and write

$$1 + \frac{1}{2} + \frac{1}{3} + \cdots + \frac{1}{p} = \frac{r}{ps}.$$

Prove that $p^3 | (r - s)$.

22. If p is a prime, prove that $\binom{n}{p} \equiv \left[\frac{n}{p}\right] \pmod{p}$. Prove also that if p^a divides at least one of $\binom{n}{p}$ and $[n/p]$, then it divides the other also.

23. Suppose that a, b and n are positive integers and n divides $a^n - b^n$. Prove that n divides $\dfrac{a^n - b^n}{a - b}$.

24. Let a, b and x_0 be positive integers, and define

$$x_n = ax_{n-1} + b \quad \text{for } n = 1, 2, 3, \ldots.$$

Prove that not all the x_n can be primes.

25. Let n, r, a denote positive integers. Prove that

$$n^2 \equiv n \pmod{10^a} \Longrightarrow n^r \equiv n \pmod{10^a} \quad \text{for all } r. \qquad (1)$$

Find all values of r such that

$$n^r \equiv n \pmod{10^a} \Longrightarrow n^2 \equiv n \pmod{10^a}. \qquad (2)$$

26. Let n, a, d be given integers with $(a, d) = 1$. Prove that there exists an integer m such that $m \equiv a \pmod{d}$ and $(m, n) = 1$.

27. Given that $[x + y] = [x] + [y]$ and $[-x - y] = [-x] + [-y]$, prove that x or y is an integer.

28. If p is prime, find formulae for the highest power of p that divides

   (a) the product $2 \cdot 4 \cdot 6 \cdots (2n)$ of the first n even integers;

   (b) the product of the first n odd integers.

29. Let p be prime and $n = a_0 + a_1p + a_2p^2 + \cdots + a_kp^k$, where a_i are integers with $0 \le a_i < p$, $a_k \ne 0$. Show that the exponent of the highest power of p that divides $n\,!$ is

$$\frac{n - (a_0 + a_1 + a_2 + \cdots + a_k)}{p - 1}.$$

30. Find all integers $n \ge 1$ such that the exponent of the highest power of 5 that divides $n\,!$ is 100.

Set IV

1. Show that $n^4 + 4$ is divisible by at least two primes for all $n > 1$.

2. Determine all positive integers n such that $n^4 + 4^n$ is divisible by two distinct primes.

3. Find the remainder obtained when the number $N = 1^1 + 2^2 + 3^3 + \cdots + 100^{100}$ is divided by 16.

4. If p is a prime number, prove that $2^p + 3^p$ is never a perfect power.

5. Find all integers n for which $n^2 + n + 41$ is a perfect square.

6. What is the least number which contains only 1's in its decimal representation and which is divisible by $33\ldots3$, where the digit 3 is repeated 100 times.

7. Which of the integers $101, 10101, 1010101, \ldots$, with alternating 0's and 1's, beginning and ending with 1, can be primes?

8. Prove that $2222^{5555} + 5555^{2222}$ is divisible by 7.

9. Prove that a product of n positive integers forming an arithmetic progression whose common difference is relatively prime to $n!$ is divisible by $n!$.

10. Find all positive integers which can be expressed as the sum of three distinct composite numbers.

11. Find all natural numbers n such that $[n^2/5]$ is a prime.

12. Let N be an even number not divisible by 10. What digit is in the ten's place of the number N^{20}?

13. Find all primes p such that $\dfrac{2^{p-1}-1}{p}$ is a perfect square.

14. Find all natural numbers n such that $n^3 - 18n^2 + 115n - 391$ is cube of some natural number m.

15. Show that $x^3 + y^3 + z^3 + t^3 = 1999$ has infinitely many integer solutions.

16. The arithmetic mean of two or more distinct primes is 27. Find the greatest of such primes.

17. Given m integers $a_1, a_2, \ldots, a_m$, show that there exist k, r with $0 \leq k < r \leq m$ such that $a_{k+1} + a_{k+2} + \cdots + a_r$ is divisible by m.

18. Suppose that 101 numbers are chosen from the set $S = \{1, 2, 3, \ldots, 200\}$. Show that two of the chosen numbers are such that one is divisible by the other.

19. Show that among any 8 composite numbers selected from the first 360 natural numbers, there will always be two which are not relatively prime.

20. Let p be an odd prime. Show that the numerator of $\sum_{k=1}^{p-1} \frac{1}{k}$ is divisible by p.

21. Let m and n be relatively prime integers with $m > n > 1$. Prove that $\frac{\log_{10} m}{\log_{10} n}$ cannot be a rational number.

22. Let a, b be positive integers such that $\frac{a+1}{b} + \frac{b+1}{a}$ is also an integer. Prove that $\gcd(a, b) \leq \sqrt{a+b}$.

23. Determine all real numbers x satisfying the equation

$$[x] + [2x] + [4x] + [8x] + [16x] + [32x] = 12345.$$

24. Let $s(N)$ be the sum of digits in the decimal representation of N. Prove that there are infinitely many values of n for which $s(2^n) > s(2^{n+1})$.

25. Determine the remainder when 3^{2^n} is divided by 2^{n+3}.

26. Prove that, for each n, there exists a positive integer A_n divisible by 2^n and whose decimal representation contains exactly n digits each of which is 1 or 2.

27. Find all integers of the form 2^n such that after deleting the first digit (from the left), we again get a power of 2.

28. Let $\{a_n\}$ be a strictly increasing sequence of positive integers such that

    (i) $a_{2n} = a_n + n$ for all $n \geq 1$, and

    (ii) if a_n is prime, then n is prime.

    Prove that $a_n = n$ for all n.

29. Find the least lcm of 20 natural numbers, not necessarily distinct, whose sum is 801.

30. Find all pairs of natural numbers whose sum is 5432 and least common multiple is 223 020.

Set V

1. Find all primes p and q such that $p^2 + 7pq + q^2$ is the square of an integer.

2. If a, b, c, d are positive integers such that $a^5 = b^4$, $c^3 = d^2$, and $c - a = 19$, find $d - b$.

3. Find the largest possible value of k for which 3^{11} is expressible as the sum of k consecutive positive integers.

4. Find the smallest positive integer whose cube ends with 888 in decimal notation.

5. Find the remainder obtained when $1997^{99} - 1997$ is divided by $1997^2 + 1998$.

6. Let n be a positive integer which is not divisible by 2 or 5. Prove that there exists an infinite number of numbers which are divisible by n and whose decimal representations contain only the digit 1.

7. Find all positive integers n such that $n + 9$, $16n + 9$ and $27n + 9$ are all perfect squares.

8. Prove that 2^n is the sum of squares of four positive integers if and only if n is even.

9. If m, n, k are positive integers, prove that

   (a) $(mn)\,!$ is divisible by $m\,!\,(n\,!)^m$,

   (b) $(mnk)\,!$ is divisible by $m\,!\,(n\,!)^m\,(k\,!)^{mn}$,

   (c) $(n^2)!$ is divisible by $(n!)^{n+1}$,

   (d) $(n\,!)\,!$ is divisible by $(n\,!)^{(n-1)!}$.

10. Let $a_1, a_2, \ldots, a_r$ be positive integers, and $n = a_1 + a_2 + \cdots + a_r$. Prove that

   (a) $N = \dfrac{n\,!}{a_1\,!\,a_2\,!\,\cdots\,a_r\,!}$ is an integer.

   (b) If, further, the gcd of $a_1, a_2, \ldots, a_r$ is d, then $N_1 = \dfrac{(n-1)\,!\,d}{a_1\,!\,a_2\,!\,\cdots\,a_r\,!}$ is an integer.

11. The difference between two prime numbers is 100. If we concatenate them (that is, juxtapose one next to another) in one order, we obtain another prime number. Find the two primes and the concatenated prime.

12. Find all the primes p and q, and even numbers $n > 2$, satisfying the equation
$$p^n + p^{n-1} + \cdots + 1 = q^2 + q + 1.$$

13. Find all pairs (x, y) of integers such that $x^3 + 11^3 = y^3$.

14. Let S denote the set of all 6-tuples (a, b, c, d, e, f) of positive integers such that $a^2 + b^2 + c^2 + d^2 + e^2 = f^2$. Consider the set
$$T = \{abcdef : (a, b, c, d, e, f) \in S\}.$$
Find the greatest common divisor of all members of T.

15. Let α and β be the roots of the quadratic equation $x^2 + mx - 1 = 0$, where m is an odd integer. Let $\lambda_n = \alpha^n + \beta^n$ for $n \geq 0$. Prove that λ_n is an integer and that $\gcd(\lambda_n, \lambda_{n+1}) = 1$.

16. Let $p_1, p_2, \ldots$ be a sequence of primes defined by $p_1 = 2$, and, for $n \geq 1$, p_{n+1} is the largest prime divisor of $p_1 p_2 \cdots p_n + 1$. (Thus, $p_2 = 3, p_3 = 7$.) Prove that $p_n \neq 5$ for any n, and also that $p_n \neq 11$ for any n.

17. Find all primes p and q such that $p(p+1) + q(q+1) = n(n+1)$ for some positive integer n.

18. For a given positive integer n, let u_n and v_n be the number of binomial coefficients $\binom{n}{k}$ which are congruent modulo 3 to 1 and 2 respectively. Prove that $u_n > v_n$. Prove also that $u_n - v_n$ is a power of 2.

19. Let p be a prime. For a given positive integer n, let u_n be the number of binomial coefficients $\binom{n}{k}$ which are not divisible by p. Prove that $u_{np+r} = (r+1)u_n$ for $0 \leq r \leq p-1$.

20. Let v_n be the number of even binomial coefficients $\binom{n}{r}$. Show that there exist an infinite number of values of n such that $v_n = 0$.

21. Find the least possible value of $a + b$, where a, b are positive integers such that 11 divides $a + 13b$ and 13 divides $a + 11b$.

22. Prove that there are infinitely many positive integers n such that $n(n+1)$ can be expressed as a sum of two positive squares in *at least* two different ways. (Here $a^2 + b^2$ and $b^2 + a^2$ are considered as the same representation.)

23. Prove that there exist infinitely many integer pairs (m, n) such that $m|(n^2+1)$ and $n|(m^2+1)$.

24. If n is divisible by an odd prime and $n > 3$, prove that $2^n + 1$ has at least two distinct prime divisors.

25. Show that the number $\left(2^{2^r}\right)^n - 1$ has at least $2r + 1$ distinct prime divisors if $n > 2^r$, unless $r = 1, n = 3$.

26. Prove that (i) 3^{n+1} divides $2^{3^n} + 1$ for all $n \geq 1$, and (ii) 3^{n+2} does not divide $2^{3^n} + 1$ for any n.

27. Let n be odd and $\gcd(a, n) = 1$. Prove that if $a^n + 1$ is divisible by n, then either $n = 1$ or $\gcd(a+1, n) > 1$.

28. Determine all positive integers n for which n^2 is a divisor of $2^n + 1$.

29. Find all integers a, b, c with $1 < a < b < c$ such that $(a-1)(b-1)(c-1)$ is a divisor of $abc - 1$.

30. The increasing sequence $1, 3, 4, 9, 10, 12, 13, \ldots$ consists of positive integers which are powers of 3 or sums of distinct distinct powers of 3. What is the 1000th term in it?

Set VI

1. Let x, y, z and n be natural numbers, and $n \geq z$. Prove that the relation $x^n + y^n = z^n$ does not hold.

2. Prove that the product of four consecutive natural numbers cannot be a perfect cube.

3. Determine all non-negative integral pairs (x, y) for which $(xy - 7)^2 = x^2 + y^2$.

4. Show that there is a natural number n such that $n\,!$ when written in decimal notation ends in exactly 1993 zeros.

5. Find the highest power of 1991 that divides the number

$$1990^{1991^{1992}} + 1992^{1991^{1990}}.$$

6. Find all natural numbers n such that the product of their digits (in decimal notation) is equal to $n^2 - 10n - 22$.

7. Let n be an integer greater than 2, and V_n be the set of integers $1 + kn$, $k = 1, 2, \ldots$. A number $m \in V_n$ is called *indecomposable* in V_n if it is not the product of two elements in V_n. Prove that there exists an element $N \in V_n$ that can be expressed as the product of elements indecomposable in V_n in more than one way.

8. Let m and n be distinct natural numbers. Suppose that, in their decimal representations, the last three digits of 1978^m are equal, respectively to the last three digits of 1978^n. Find m, n such that $m + n$ has the least value.

9. Show that there exist infinitely many pairs (a, b) of positive integers such that

   (i) $ab(a + b)$ is not divisible by 7,
   (ii) $(a + b)^7 - a^7 - b^7$ is divisible by 7^7.

10. Let a, b, c and d be odd integers such that $0 < a < b < c < d$ and $ad = bc$. Prove that if $a + d = 2^k$ and $b + c = 2^m$ for some positive integers k and m, then $a = 1$. Determine all such integers a, b, c, d.

11. Show that there exist infinitely many sets of 1983 consecutive positive integers each of which is divisible by some number of the form a^{1983} with $a \geq 2$.

12. Each of the numbers $x_1, x_2, \ldots, x_n$ equals 1 or -1. If

$$\sum_{i=1}^{n} x_i x_{i+1} x_{i+2} x_{i+3} = 0,$$

where $x_{n+1} = x_1$, $x_{n+2} = x_2$ and $x_{n+3} = x_3$, prove that n is divisible by 4.

13. Factor the number $5^{1985} - 1$ into a product of three integers each of which is larger than 5^{100}.

14. If $(1+x+x^2+x^3+x^4)^{496} = a_0 + a_1x + \cdots + a_{1984}x^{1984}$,

    (a) determine the gcd of the coefficients $a_3, a_8, a_{13}, \ldots, a_{1983}$;

    (b) show that $10^{347} > a_{992} > 10^{340}$.

15. Let k be a positive integer, $a_0 = 0$ and, for $n = 0, 1, 2, \ldots$,

$$a_{n+1} = k(a_n+1) + (k+1)a_n + 2\sqrt{k(k+1)a_n(a_n+1)}.$$

    Prove that a_n is a positive integer for $n = 1, 2, 3, \ldots$.

16. Let d be any positive integer not equal to 2, 5 or 13. Show that one can find distinct a, b in the set $\{2, 5, 13, d\}$ such that $ab - 1$ is not a square.

17. Let p be a prime. Prove that if $f(n) = n^2 + n + p$ is prime for all integers n such that $0 \le n \le \sqrt{p/3}$, then $f(n)$ is prime for all integers n such that $0 \le n \le p-2$.

18. (a) Let a and b be positive integers such that $ab+1$ divides a^2+b^2. Show that $\dfrac{a^2+b^2}{ab+1}$ is the square of an integer.

    (b) Let k be an integer > 1. Show that we can find any number of pairs (a, b) of positive integers such that $\dfrac{a^2+b^2}{ab+1} = k^2$. also, show that $a \ge k$ and $b \le a^3$ for all such pairs.

19. Let $n > 6$ be an integer and $a_1, a_2, \ldots, a_k$ be all the natural numbers less than n and relatively prime to n. If

$$a_2 - a_1 = a_3 - a_2 = \cdots = a_k - a_{k-1} > 0,$$

    prove that n must be either a prime or a power of 2.

20. Determine all ordered pairs (m, n) of positive integers such that $\dfrac{n^3+1}{mn-1}$ is an integer.

21. Let p be a prime. Show that the congruence $x^2 + x + 3 \equiv 0 \pmod{p}$ is solvable if and only if $x^2 + x + 25 \equiv 0 \pmod{p}$ is solvable.

22. Let the sum of reciprocals of integers $1, 2, 3, \ldots, 100$ be written as a/b, where a, b are positive integers which are relatively prime. Prove that ab is not divisible by 5.

23. Let p be a prime greater than 3. If the sum of reciprocals of $1, 2, \ldots, p-1$ is m/n, where m and n are integers, prove that $p^2|m$.

24. Show that the equation

$$x^4 + y^4 + z^4 - 2x^2y^2 - 2y^2z^2 - 2z^2x^2 = 24$$

has no solution in integers.

25. If x and y are positive integers such that xy divides $x^2 + y^2 - x$, show that x is a perfect square.

26. If n is a natural number for which $4^n + 2^n + 1$ is a prime, show that n is a power of 3.

27. Find all natural numbers n which are sums of two odd composite natural numbers.

28. Show that there exists a set S of 2000 distinct positive integers such that each element of S is less than 10^5 and no three of S form an arithmetic progression.

29. Note that $12^2 = 144$ ends in two 4's. Determine the length of the longest string of equal non-zero digits in which the square of an integer can end.

30. For an integer $k \geq 2$, define $f : \mathbf{N} \longrightarrow \mathbf{N}$ by

$$f(n) = \left[\left(n + n^{1/k}\right)^{1/k}\right] + n,$$

where $[x]$ denotes the integral part of x. Prove that the range of f omits exactly those natural numbers which are k-th powers.

Set VII

1. Let n be a positive integer. Prove that if $2 + 2\sqrt{28n^2 + 1}$ is an integer, then it is a square.

2. Find all pairs (n, k) of positive integers such that the congruence

$$a^k \equiv 0 \text{ or } 1 \pmod{n^k}$$

holds for every integer a.

3. Determine all integral solutions of the equation $x^2 + y^2 + z^2 = x^2y^2$.

4. Determine all integers k such that the equation $x^2 + y^2 + z^2 = kxyz$ has a non-trivial solution in integers.

5. Let r be an integer greater than 1 and $2 \leq n \leq (2^r - 1)^2$. If $x_1, \ldots, x_n$ are integers such that

$$x_1^r + x_2^r + \cdots + x_n^r = x_1^r x_2^r \cdots x_n^r, \qquad x_i \geq 0 \text{ for all } i.$$

Prove that $x_1 = x_2 = \cdots = x_n = 0$.

6. Show that the equation

$$x_1^4 + x_2^4 + \cdots + x_{14}^4 = 1599$$

has no solution in integers.

7. Let n be a positive integer. Determine whether or not there are distinct positive integers $x_1, \ldots, x_n$ such that

$$x_1^2 + x_2^2 + \cdots + x_n^2 = y^3 \quad \text{and} \quad x_1^3 + x_2^3 + \cdots + x_n^3 = z^2,$$

simultaneously, for integers y, z.

8. An integer n will be called *good* if we can write

$$n = a_1 + a_2 + \cdots + a_k,$$

where $a_1, a_2, \ldots, a_k$ are positive integers (not necessarily distinct) such that the sum of their reciprocals is 1. Given the information that the integers 33 through 73 are good, prove that every $n \geq 33$ is good.

9. Prove that the cube roots of three distinct prime numbers cannot be three terms (not necessarily consecutive) of an arithmetic progression.

10. Show that the cube roots of three distinct prime numbers cannot be three terms (not necessarily consecutive) of a geometric nor of a harmonic progression.

11. Prove that, for every positive integer n there is a finite set S_n of positive integers such that it has n elements and the geometric mean of every subset of S_n is an integer. Is there an infinite set S of positive integers such that the geometric mean of any finite subset of S is an integer?

12. Find the least value of $n > 1$ such that the root-mean-square of the first n natural numbers is an integer.

13. Find all values of n such that $1 < n < 100$ and the standard deviation of the first n natural numbers is an integer.

14. (a) Show that there exists an odd positive integer k such that $2^n + k$ is composite for all positive integers n.
    (b) Show that there exists a positive integer k such that $k \cdot 2^n + 1$ is composite for all positive integers n.

15. Do there exist 14 consecutive positive integers each of which is divisible by one or more primes p from the interval $2 \le p \le 11$?

16. Do there exist 21 consecutive positive integers each of which is divisible by one or more primes p from the interval $2 \le p \le 13$?

17. The nth triangular number t_n is given by $t_n = n(n+1)/2$. Prove that
    (a) n is an even positive integer such that t_n is a perfect square if and only if there exist integers x and y satisfying the equation $x^2 - 2y^2 = 1$ and $n = 2y^2$;
    (b) n is an odd positive integer such that t_n is a perfect square if and only if there exist integers x and y satisfying the equation $x^2 - 2y^2 = -1$ and $n = 2y^2 - 1$;
    (c) there exist an infinite number of even integers n, as well as odd integers n, such that t_n is a square.

18. Determine all positive integers n such that $n!$ is divisible by the sum $1 + 2 + \cdots + n$.

19. If a and b are positive integers such that $a|b^2$, $b^2|a^3$, $a^3|b^4$, $b^4|a^5, \ldots$, prove that $a = b$.

20. Prove that if $n > 1$, then $s_n = \sum_{j=1}^{n} 1/j$ is not an integer.

21. Prove that if $n > m \ge 1$, then $s_n = \sum_{j=m}^{n} 1/j$ is not an integer.

22. Let d be a positive integer, and

$$s_n = \frac{1}{1+d} + \frac{1}{1+2d} + \cdots + \frac{1}{1+nd}.$$

Assuming Bertrand's theorem, which states that given an integer $n > 1$, we can find a prime p such that $n \le p < 2n$, prove that s_n is not an integer for any $n \ge 1$.

23. Prove that n is expressible as a sum of consecutive positive integers if and only if n is not a power of 2.

24. Let $k \geq 3$ be a fixed integer. Find all finite sequences $a_1, a_2, \ldots, a_k$ of positive integers such that the sum of any three terms in it is divisible by each of these three terms.

25. Prove that there is no non-constant polynomial $f(x)$ with integer coefficients such that $f(x)$ is prime for every positive integer x.

26. Find all triples (a, b, c) of non-zero integers satisfying the following conditions:

$$a \equiv b \pmod{|c|}, \quad b \equiv c \pmod{|a|}, \quad c \equiv a \pmod{|b|},$$
$$\gcd(a, b, c) = 1, \quad |a| \leq |b| \leq c.$$

27. Find the smallest positive integer n such that $\phi(x) = n$ has no solution; exactly two solutions; exactly three solutions; exactly four solutions.

28. Find all solutions x of $\phi(x) = 24$.

29. Suppose that $I(x)$ denotes the integer nearest to x, where we choose the larger one if there are two such integers. Prove that for any positive integer n, we have

$$\sum_{k=1}^{\infty} I(n/2^k) = n.$$

30. Let $f(x) = x - [x] - \frac{1}{2}$. Prove that

$$\left|\sum_{n=1}^{m} f\left(2^n x + \frac{1}{2}\right)\right| \leq 1 \quad \text{for all } m \geq 1 \text{ and all real } x.$$

Set VIII

1. If n is a positive integer, prove that

$$\left[\sqrt{n} + \sqrt{n+1}\right] = \left[\sqrt{4n+1}\right] = \left[\sqrt{4n+2}\right] = \left[\sqrt{4n+3}\right].$$

Determine all integers r such that

$$\left[\sqrt{n} + \sqrt{n+1}\right] = \left[\sqrt{4n+r}\right] \quad \text{for all } n \geq 1.$$

2. If n is a positive integer, prove that

$$\left[\frac{8n+13}{25}\right]-\left[\frac{n-12-[(n-17)/25]}{3}\right]$$

is independent of n.

3. Prove that, for $x>0$, $\sum_{n\le x}\left[\sqrt{\frac{x}{n}}\right]=\sum_{n\le\sqrt{x}}\left[\frac{x}{n^2}\right]$.

4. Prove that $\sum_{k=1}^{n}(2k-1)\left[\frac{n}{k}\right]=\sum_{k=1}^{n}\left[\frac{n}{k}\right]^2$ for all positive integers n.

5. Prove that (i) $\sum_{k=1}^{n}\left[\frac{k}{2}\right]=\left[\frac{n^2}{4}\right]$ and (ii) $\sum_{k=1}^{n}\left[\frac{k}{3}\right]=\left[\frac{n(n-1)}{6}\right]$, for all positive integers n.

6. If $a=1,2,\ldots,7$, prove that there exists an integer b (depending on a) such that

$$\sum_{k=1}^{n}\left[\frac{k}{a}\right]=\left[\frac{(2n+b)^2}{8a}\right] \quad \text{for all positive integers } n.$$

7. Show that $\left[(1+\sqrt{3})^{2n}\right]+1$ and $\left[(1+\sqrt{3})^{2n+1}\right]$ are both divisible by 2^{n+1}. Is this the highest power of 2 dividing both the numbers?

8. If 2^k-1 is prime ($k>1$), show that $n=2^{k-1}(2^k-1)$ is perfect. Show also that every even perfect number is of this form.

9. Let n be an even perfect number. Show that $n\equiv 6 \pmod{10}$ or $n\equiv 28 \pmod{100}$. Show also that if $n\equiv 6 \pmod{100}$, then $n=6$.

10. Let n be an even perfect number, and $n>6$. Prove that $n\equiv 1 \pmod 9$ and $n\equiv 4 \pmod 6$.

11. Let n be an even perfect number, and $n>6$. Prove that n can be expressed as a sum of consecutive odd cubes.

12. Prove that no proper divisor of a perfect number can be perfect.

13. Prove that a perfect square cannot be a perfect number.

14. Let $H(n)$ be the harmonic mean of the positive divisors of n. Prove that $H(n)$ is an integer if n is perfect. Does there exist numbers n for which $H(n)$ is an integer, but n is not perfect?

15. (a) Prove that no power of a prime can be a perfect number.

    (b) If n is an odd perfect number, prove that n has at least three distinct prime divisors.

16. If n is an odd perfect number, prove that n can be expressed as $n = p_1^{k_1} p_2^{k_2} \cdots p_r^{k_r}$, where the p_i are distinct primes, $p_1 \equiv k_1 \equiv 1 \pmod 4$ and $k_2, \ldots, k_r$ are even.

    In the following problems, u_n denotes the n-th Fibonacci number, defined by $u_1 = u_2 = 1$, $u_{n+1} = u_n + u_{n-1}$ for $n > 2$.

17. Prove that

    (a) $u_{m+n} = u_{m-1}u_n + u_m u_{n+1}$ for all $m \geq 2, n \geq 1$,

    (b) $u_{n+2}^2 - u_n^2 = u_{2n+2}$.

18. Prove that

    (a) $2^{n-1}u_n \equiv n \pmod 5$ for all $n \geq 1$,

    (b) $u_1^2 + u_2^2 + \cdots + u_n^2 = u_n u_{n+1}$ for $n \geq 1$,

19. Prove that, for $m, n \geq 1$, we have

    (a) $u_{5n+2} > 10^n$ for all $n \geq 1$,

    (b) u_{mn} is divisible by u_m,

    (c) u_{3n} is even, u_{4n} is divisible by 3, and u_{5n} is divisible by 5.

20. Prove that

    (a) $u_m u_n - u_{m+1}u_{n-1} = (-1)^{n-1}u_{m-n+1}$ for $m \geq n \geq 2$,

    (b) $u_n^2 - u_{n+1}u_{n-1} = (-1)^{n-1}$ for $n \geq 2$,

    (c) if m, n and $m - n$ are positive integers, then a common divisor of u_m and u_n is also a divisor of u_{m-n}.

21. Prove that, for all positive integers m, n,

    (a) $\gcd(u_m, u_n) = u_d$ where $d = \gcd(m, n)$,

    (b) $u_m | u_n$ if and only if $m | n$.

22. Prove that there is no positive integer n for which $u_1+u_2+\cdots+u_{3n} = 16\,!$.

23. Prove that, for any positive integer n, we can find n consecutive composite Fibonacci numbers.

24. Prove that any positive integer N can be expressed as a sum of distinct Fibonacci numbers, no two of which are consecutive.

25. Let $\alpha = \frac{1}{2}(1+\sqrt{5})$ and $\beta = \frac{1}{2}(1-\sqrt{5})$. Prove that, for all $n \geq 1$,

    (a) $u_n = \dfrac{\alpha^n - \beta^n}{\sqrt{5}}$,

    (b) u_n is the nearest integer to $\dfrac{\alpha^n}{\sqrt{5}}$.

26. Let p be a prime and $p > 5$. Prove that either $p|u_{p-1}$ or $p|u_{p+1}$.

27. Prove that the product $u_n u_{n+1} u_{n+2} u_{n+3}$ of any four consecutive Fibonacci numbers is equal to the area of a Pythagorean triangle.

28. If m is an even positive integer, prove that the sum of any $2m$ consecutive Fibonacci numbers is divisible by u_m. Also determine all odd integers m with this property. Deduce that the sum of any 24 consecutive Fibonacci numbers is divisible by 24.

29. Determine all integers n such that $u_n + 1$ is prime.

30. Prove that, for $n \geq 2$,

$$u_n = \binom{n-1}{0} + \binom{n-2}{1} + \binom{n-3}{2} + \cdots,$$

    where we use the convention that $\binom{m}{0} = 1$, and $\binom{m}{r} = 0$ when $m < r$ or $m < 0$.

Set IX

1. Let $M_n = 2^n - 1$ be the nth Mersenne number. Suppose that p and $q = 2p+1$ are primes. Prove that (i) q divides M_p if $q \equiv 1$ or $7 \pmod 8$, and (ii) q divides $M_p + 2$ if $q \equiv 3$ or $5 \pmod 8$. Deduce that M_{11} and M_{23} are not primes.

2. If p is an odd prime, prove that any prime divisor of M_p is of the form $2kp + 1$. Deduce that M_{13} is prime.

3. If p is an odd prime, prove that any prime divisor q of M_p is of the form $q = 8m \pm 1$.

4. Prove that the Mersenne number M_{29} is composite by finding its smallest prime divisor.

5. (a) Prove that any two Fermat numbers are relatively prime, that is, $\gcd(F_m, F_n) = 1$ for $m > n \geq 0$. Deduce that the number of primes is infinite.

   (b) Use the sequence $\{a_n\}$, defined by

$$a_1 = 2, \quad a_{n+1} = a_n^2 - a_n + 1, \quad \text{for } n \geq 1,$$

   to show that the number of primes is infinite.

6. Prove that if p is prime and $p > 3$, then,

$$3^{(p-1)/2} \equiv (-1)^{[(p+1)/6]} \pmod{p}.$$

7. Prove that, for $n \geq 1$, the Fermat number F_n is prime if and only if

$$3^{(F_n-1)/2} \equiv -1 \pmod{F_n}.$$

8. Any prime divisor p of the Fermat number $F_n = 2^{2^n} + 1$, where $n \geq 2$, is of the form $p = k \cdot 2^{n+2} + 1$. Deduce that $F_4 = 65537$ is prime.

9. For $n \geq 1$, prove the following:

   (a) There are infinitely many composite numbers of the form $2^{2^n} + 3$.

   (b) Each of numbers $2^{2^n} + 5$ is composite.

10. If $2^n \equiv 2 \pmod{n}$ and n is composite, then n is said to be a pseudo-prime. Show that every Fermat number is either a prime or a pseudo-prime.

11. If p and q are both primes and $q | (2^p + 1)$, prove that either $q = 3$ or $q = 2kp + 1$ for some integer k. Use this to determine the smallest prime divisor $q > 3$ of $2^{29} + 1$.

12. If p is a prime, prove that the order of 2 modulo p is a power of 2 if and only if p divides a Fermat number, F_k. also, show that there is no Fermat number that is divisible by 7, 13 or 97.

13. For any prime $p > 3$, prove that $(2^p + 1)/3$ has a prime divisor greater than p.

14. Let p be an odd prime. Prove that

    (a) p can be expressed as the difference of two squares,

    (b) p can be expressed as the difference of two cubes if and only if it is of the form $6t_n + 1$, where t_n is a triangular number,

    (c) any Fermat prime can be expressed as the difference of two squares, but not of two cubes.

15. (a) If u and v are real numbers, and m is a positive integer, prove that

$$u^{2m} + u^{2m-1}v + u^{2m-2}v^2 + \cdots + v^{2m} \geq \frac{1}{2}u^{2m}.$$

    (b) If p and q are odd primes and if $q = x^p - y^p$, where x, y are integers, prove that $q \equiv 1 \pmod{p}$.

16. If a Fermat prime F_n can be expressed as $F_n = x^k - y^k$, where x, y are integers, prove that $k = 2$.

17. Prove that a Carmichael number is necessarily odd.

18. (a) Let $n = 2p$, where p is an odd prime. Prove that $a^{n-1} \equiv a \pmod{n}$ for any a.

    (b) Does there exist an odd number n with the property that $a^{n-1} \equiv a \pmod{n}$ for every a?

    (c) If $n = 195$, prove that $a^{n-2} \equiv a \pmod{n}$ for any a.

19. Show that 341 and 645 are pseudo-primes, but not Carmichael numbers.

20. Show that $2 \cdot 73 \cdot 1103$ is an even pseudo-prime.

21. Show that there exist an infinite number of pseudo-primes.

22. (a) Let p be a prime. Prove that p^2 is a pseudo-prime if and only if $2^{p-1} \equiv 1 \pmod{p^2}$.

(b) Show that there are positive integers n such that n^2 is a pseudoprime.

23. (a) Let $a > 1$ and m be a composite number such that $a^m \equiv a \pmod{m}$. If m is relatively prime to $a - 1$, show that there exist an infinite number of composite numbers n such that $a^n \equiv a \pmod{n}$.

 (b) Show that there exist an infinite number of composite numbers n such that $3^n \equiv 3 \pmod{n}$.

24. Let $n = p_1 p_2 \cdots p_k$, where the p_i are distinct primes and $k > 1$. If $p_i - 1$ divides $n - 1$ for each i, prove that $a^n \equiv a \pmod{n}$ for all positive integers a.

25. Suppose that n is a composite number such that $a^{n-1} \equiv 1 \pmod{n}$ whenever a is relatively prime to n. Prove that n is product of distinct primes.

26. Let $n = p_1 p_2 \cdots p_k$, where the p_i are distinct primes. Suppose also that $a^{n-1} \equiv 1 \pmod{n}$ whenever a is relatively prime to n. Prove that $p_i - 1$ divides $n - 1$ for each i.

27. Prove that for a composite number n, the following two statements are equivalent:

 (i) $a^n \equiv a \pmod{n}$ for all positive integers a;

 (ii) $a^{n-1} \equiv 1 \pmod{n}$ for all positive integers a which are relatively prime to n.

28. Prove that 561, 1105, 1729 and 2465 are Carmichael numbers.

29. Prove that if $6k + 1$, $12k + 1$ and $18k + 1$ are primes, then

$$n = (6k + 1)(12k + 1)(18k + 1)$$

is a Carmichael number. Verify that 1729 is number of this form.

30. Suppose that n is a composite number such that $a^{n-1} \equiv 1 \pmod{n}$ whenever a is relatively prime to n. Prove that n is product of at least three primes.

Set X

1. Recall that an integer $n > 1$ is highly composite if it has more divisors than all positive integers less than it. Any $n > 1$ can be expressed as

$$n = 2^{n_1}3^{n_2}5^{n_3}\cdots p_r^{n_r},$$

where p_i is the ith prime, $n_i \geq 0$ and $n_r > 0$. Prove that if n is highly composite, then $n_1 \geq n_2 \geq \cdots \geq n_r$. Using Bertrand's theorem, or otherwise, show that the last exponent n_r is either 1 or 2. Determine those highly composite numbers for which $n_r = 2$.

2. Let $\gcd(a, b) = 1$ and c be a positive integer. Prove that there is an integer x such that $\gcd(a + bx, c) = 1$.

3. Suppose that $p = 12k + 1$ and $q = 24k + 1$ are primes for some natural number k. If $n = pq$, prove that $2^n \equiv 2 \pmod{n}$ and $3^n \equiv 3 \pmod{n}$.

4. (a) Let p be an odd prime and t be a positive integer not divisible by p. Prove that if p can be represented as $p = x^2 + ty^2$ with positive integers x and y, then $(-t)^{(p-1)/2} \equiv 1 \pmod{p}$, that is, $-t$ is a quadratic residue of p.

   (b) Prove that a prime p can be expressed as $p = a^2 + 2b^2$ for some positive integers a and b, if and only if p is of the form $8k + 1$ or $8k + 3$.

5. (a) Prove that a prime p can be represented as $p = a^2 + 3b^2$ with positive integers a and b, if and only if it is of the form $6k + 1$.

   (b) Prove that a prime p can be simultaneously expressed in the forms $x^2 + y^2$, $u^2 + 2v^2$, and $r^2 + 3s^2$, if and only if $p = 24k + 1$ for some integer k.

6. Prove that there exist an infinite number of primes

   (a) of the form $8k + 3$, and

   (b) of the form $6k + 1$.

7. Prove the following generalisation of Euler's theorem:

$$a^n \equiv a^{n-\phi(n)} \pmod{n}$$

for all integers a.

8. Let n be an integer greater than 1. Prove that n is prime if and only if there is a positive integer a such that

$$a^{n-1} \equiv 1 \pmod{n} \quad \text{but} \quad a^d \not\equiv 1 \pmod{n}$$

for every proper divisor d of $n - 1$.

9. (a) Let p be a prime. If $\gcd(a, p) = 1 = \gcd(n, p - 1)$, prove that the congruence $x^n \equiv a \pmod{p}$ has a solution, unique modulo p.
   (b) Let p be a prime of the form $3k + 2$. Show that the congruence $x^3 \equiv a \pmod{p}$ has a solution for $1 \leq a \leq p - 1$. also, show that if p is of the form $3k + 1$, then this congruence has a solution for only one-third of such values of a.

10. (a) Let $n \geq 3$ and k be odd. Prove that the congruence $x^2 \equiv k \pmod{2^n}$ has a solution if and only if $k \equiv 1 \pmod{8}$.
    (b) Let $n \geq 3$ and $k \equiv 1 \pmod{8}$. Prove that the congruence $x^2 \equiv k \pmod{2^n}$ has exactly four incongruent solutions.

11. Let d be a positive integer which is not a perfect square. It is known that, for any such d, the equation $x^2 - dy^2 = 1$ has a solution in positive integers x, y.
    (a) Prove that $x^2 - dy^2 = -1$ has no solution if $d \equiv 3 \pmod{4}$.
    (b) Prove that if $x^2 - dy^2 = N$ has one solution, it has infinitely many solutions.
    (c) If k is any positive integer, prove that there are infinitely many solutions in integers of $x^2 - dy^2 = 1$ with $k|y$.

12. Determine all the Pythagorean triples (x, y, z) forming (a) an arithmetic progression, (b) a geometric progression.

13. If x, y, z is a primitive Pythagorean triple, prove that $x + y$ and $x - y$ are congruent modulo 8 to 1 or -1.

14. (a) Prove that every positive integer n can be expressed in the form $n = x^2 + y^2 - z^2$.
    (b) Prove that if x, y, z are integers such that $x^2 + y^2 = z^2$, then xy is divisible by 12, and xyz is divisible by 60.
    (c) Prove that if x, y, z are non-zero integers such that $\dfrac{1}{x^2} + \dfrac{1}{y^2} = \dfrac{1}{z^2}$, then xy is divisible by 60.

15. Consider the collection of primitive Pythagorean triples x, y, z, where x, z are odd and y is even. Prove that there exist infinitely many such triples (x, y, z) in which

    (a) the even member y is a perfect square,

    (b) the odd member x is a perfect square,

    (c) the odd member z is a perfect square.

16. Prove that there is no Pythagorean triple (x, y, z) in which both x and y are perfect squares.

17. Let (x, y, z) be a primitive Pythagorean triple in which z is a perfect square. Prove that neither x nor y can be a perfect square.

18. Consider the collection of primitive Pythagorean triples (x, y, z) where x, z are odd and y is even. Prove that, for a given n, there exist at least n such triples having

    (a) the same value for x,

    (b) the same value for y.

19. Prove that there exits an infinite number of primitive Pythagorean triples (x, y, z) in which x and y are consecutive positive integers. Give three such triples.

20. (a) Prove that there exist infinitely many primitive Pythagorean triples (x, y, z) in which x and y are consecutive triangular numbers.

    (b) Deduce from (a) that there exist infinitely many triangular numbers that are perfect squares. Obtain three such triangular numbers.

21. Prove that we can find a sequence $a_1, a_2, a_3, \ldots$ of positive integers such that, for all $n \geq 2$,

$$a_1^2 + a_2^2 + \cdots + a_n^2 = \text{the square of an integer.}$$

22. Prove that the radius of the incircle of a Pythagorean triangle is always an integer.

23. For a given positive integer n, show that there is a Pythagorean triangle the radius of whose incircle is n. Show also that if n is an odd prime, then there are precisely three Pythagorean triangles with this property.

24. Find all Pythagorean triangles whose areas are equal to their perimeters.

25. Prove that

(a) the Diophantine equation $x^4 - y^4 = z^2$ has no solution with $yz \neq 0$,

(b) there are no positive integers a, b, such that both $a^2 + b^2$ and $a^2 - b^2$ are squares of integers, and $a \neq b$,

(c) there is no Pythagorean triangle whose area is the square of an integer.

26. Assume, for convenience, that in a primitive Pythagorean triple (x, y, z) of positive integers, the first component x is odd and the second component y is even. Suppose that (x, y, z) and (a, b, c) are two such primitive triples. Let $\gcd(c, z) = 1$ and $w = cz$. Prove that they can be combined to form two triples (u_1, v_1, w) and (u_2, v_2, w) of the same type and having the same third component w.

27. Assume, as before, that in a primitive Pythagorean triple, the first component is odd and the second component is even. Let (x, y, z) be such a triple, p be a prime congruent to 1 modulo 4, $\gcd(z, p) = 1$ and $w = pz$.

(a) Prove that (x, y, z) *generates* two triples (u_1, v_1, w) and (u_2, v_2, w) of the same type and having the same third component w.

(b) also, show that every such triple (u, v, w), with this w, corresponds to a triple (x, y, z) which generates (u, v, w).

28. Assume, as in the previous problems, that in a primitive Pythagorean triple, the first component is odd and the second component is even. Suppose that there are exactly m such triples (x_i, y_i, z), $i = 1, \ldots, m$, having the same third component z. Let p be a prime of the form $4k+1$, relatively prime to z, and $w = pz$. Prove that there exist exactly $2m$ such primitive triples having the same third component w.

29. Let N be the product of k distinct primes each of which is congruent to 1 modulo 4. Prove that there exist exactly 2^{k-1} primitive Pythagorean triples (x_i, y_i, N) whose first components x_i are odd and second components y_i are even, and have the same third component N.

30. As in the previous problem, let N be the product of k distinct primes each of which is congruent to 1 modulo 4.

(a) Determine the number of Pythagorean triples (x, y, N) having the same third component N.

(b) Find all Pythagorean triples $(x, y, 1105)$.

THREE

Solutions

Set I

1. The nth *triangular number* t_n is the sum of the first n natural numbers. Thus, $t_n = n(n+1)/2$ for $n = 1, 2, \ldots$. If the nth triangular number t_n is a perfect square, prove that $t_{4n(n+1)}$ is also a perfect square. Deduce that there are an infinite number of triangular numbers which are perfect squares.

   Solution Let $t_n = a^2$. Then, $8a^2 = 4n(n+1)$, $8a^2 + 1 = (2n+1)^2$, and hence, $8a^2(8a^2+1)/2 = (2a)^2(2n+1)^2$. Thus, if $m = 8a^2 = 4n(n+1)$, then t_m is a perfect square.

   Since $t_1 = 1$ is a square, we see that t_8 is a square. Repeated application of this shows that there are an infinite number of values of m such that t_m is a square.

2. Find three consecutive triangular numbers t_n, t_{n+1}, t_{n+2}, whose product is a perfect square. Also show that there exists an infinite number of triples of consecutive triangular numbers whose product is a perfect square.

   Solution We have

$$t_n t_{n+1} t_{n+2} = \frac{n(n+1)^2(n+2)^2(n+3)}{8} = \frac{n(n+3)}{2} t_{n+1}^2.$$

   If $n = 3m$, then $\frac{1}{2}n(n+3) = \frac{1}{2}3m(3m+3) = 9t_m$. Hence, if $t_m = a^2$, then

$$t_n t_{n+1} t_{n+2} = 9a^2 t_{n+1}^2 = (3at_{n+1})^2.$$

Since $t_1 = 1^2$ and $t_8 = 6^2$, we get

$$t_3t_4t_5 = (3 \cdot 10)^2, \quad t_{24}t_{25}t_{26} = (18 \cdot t_{25})^2.$$

It has been proved above that there exist an infinite number of values of m such that t_m is perfect square. Correspondingly, we get an infinite number of values of $n = 3m$ such that $t_n t_{n+1} t_{n+2}$ is a perfect square.

3. Verify that 10 is the only triangular number that can be written as the sum of two consecutive odd squares.

Solution Clearly, 10 can be written as $10 = 1^2 + 3^2$. Suppose that $t_n = (2k-1)^2 + (2k+1)^2$ for some integer k. Then, $n(n+1)/2 = 2(4k^2+1)$, $4n(n+1) = 16(4k^2+1)$, and hence,

$$(2n+1)^2 = 64k^2 + 17, \quad (2n+1+8k)(2n+1-8k) = 17.$$

This implies that $2n+1-8k = 1$ and $2n+1+8k = 17$. Adding, $2(2n+1) = 18$, that is, $n = 4$, $t_n = 10$.

4. Let n be a positive integer. If $a_1, \ldots, a_n$ are integers and the equation

$$x^n + a_1x^{n-1} + a_2x^{n-2} + \cdots + a_n = 0$$

has a rational root, prove that the root must be an integer.

Solution Suppose that $r = a/b$ is a root, where a and b are integers. We may suppose that $\gcd(a, b) = 1$ and $b \geq 1$; otherwise, we can cancel the common factors. We get

$$\frac{a^n}{b^n} + a_1\frac{a^{n-1}}{b^{n-1}} + a_2\frac{a^{n-2}}{b^{n-2}} + \cdots + a_n = 0.$$

$$a^n + a_1a^{n-1}b + a_1a^{n-2}b^2 + a_nb^n = 0. \qquad (*)$$

If $b > 1$, then b has a prime divisor p. We see from (*) that p divides a^n and hence a. Thus, p is a common divisor of a and b, which is impossible by our assumption. It follows that $r = a$, an integer.

5. If $(a, 4) = 2 = (b, 4)$, prove that $(a+b, 4) = 4$.

Solution Since $(a, 4) = 2$, we have $a = 2a_1$, and so $2 = (2a_1, 4) = 2(a_1, 2)$. This shows that $(a_1, 2) = 1$. Similarly, $b = 2b_1$ with $(b_1, 2) = 1$. Thus, a_1 and b_1 are odd, and so $a + b = 2(a_1 + b_1)$ is divisible by 4 since $a_1 + b_1$ is even.

6. If a, b, c are positive integers such that $(a, b, c)[a, b, c] = abc$, prove that $(a, b) = (b, c) = (a, c) = 1$.

Solution Let $a = \prod_i p_i^{a_i}$, $b = \prod_i p_i^{b_i}$ and $c = \prod_i p_i^{c_i}$, where a_i, b_i, c_i are non-negative integers and p_i are distinct primes. The given equality becomes

$$\prod_i p_i^{\min(a_i,b_i,c_i)} \prod_i p_i^{\max(a_i,b_i,c_i)} = \prod_i p_i^{a_i+b_i+c_i}.$$

Consequently,

$$\min(a_i, b_i, c_i) + \max(a_i, b_i, c_i) = a_i + b_i + c_i \quad \text{for each } i.$$

If, for some i, $a_i \leq b_i \leq c_i$, then this implies that $b_i = 0$, and hence, $a_i = 0$. Thus, two of a_i, b_i, c_i must be zero, for each i. If $(a, b) > 1$ then, for some i, $p_i|a$ and $p_i|b$, that is, $a_i > 0$ and $b_i > 0$. This shows that $(a, b) = 1$. Similarly, $(b, c) = 1 = (a, c)$.

Another solution: The product of two numbers is equal to the product of their g.c.d and l.c.m. Thus,

$$ab = d[a, b], \quad [a, b]c = x\,[[a, b], c],$$

where $d = (a, b)$ and $x = ([a, b], c)$. Since $[[a, b], c] = [a, b, c]$, we get

$$(a, b, c)[a, b, c] = abc = d[a, b]c = dx[a, b, c],$$

and so $(a, b, c) = dx$. Since $(a, b, c) = ((a, b), c) = (d, c)$, this gives $(d, c) = dx$. Thus,

$$d|(d, c), \quad (d, c) = d, \quad dx = (d, c) = d, \quad x = 1.$$

Hence, $[a, b]$ and c are relatively prime. A common divisor of c and a is also a common divisor of c and $[a, b]$. It follows that $(c, a) = 1$. So, $dx = (a, b, c) = 1$, $d = 1$.

7. Prove that $[a, b, c](ab, bc, ca) = abc$ for all positive integers a, b, c.

Solution Consider, as above, the standard representations of a, b, c. Then,

$$ab = \prod_i p_i^{a_i+b_i}, \quad bc = \prod_i p_i^{b_i+c_i}, \quad ca = \prod_i p_i^{c_i+a_i}.$$

Hence, $[a, b, c](ab, bc, ca) = \prod_i p_i^{n_i}$, where

$$n_i = \max(a_i, b_i, c_i) + \min(a_i + b_i, b_i + c_i, c_i + a_i).$$

If, for some i, $\max(a_i, b_i, c_i) = c_i$, then,

$$\min(a_i + b_i, b_i + c_i, c_i + a_i) = a_i + b_i,$$

and hence $n_i = a_i + b_i + c_i$. It follows that

$$[a, b, c](ab, bc, ca) = \prod_i p_i^{a_i+b_i+c_i} = abc.$$

8. If m, n are positive integers such that $3m + n = 3[m, n] + (m, n)$, prove that n divides m.

Solution Let $d = (m, n)$, $m = m_1 d$ and $n = n_1 d$. Then, $[m, n] = mn/d = m_1 n_1 d$, and hence,

$$(3m_1 + n_1)d = 3m_1 n_1 d + d, \quad (3m_1 - 1)(n_1 - 1) = 0.$$

Since $3m_1 \neq 1$, we have $n_1 = 1$, $n = d$ which divides m.

9. Examine whether the following statements are true or false. If true, prove the result, and if false, give a counterexample:

(a) If $(a, b) = 1$, then $(a^2, ab, b^2) = 1$.

(b) If $(a, b) = 1$, then $(a + b, a^2 - ab + b^2)$ is either 1 or 3.

(c) If $b|(a^2 + 1)$, then $b|a^4 + 1$.

(d) Every positive integer can be written in the form $p + a^2$, where p is a prime or 1, and $a \geq 0$.

(e) If p is prime and $p > 3$, then $p^2 + 2$ is composite.

Solution (a) If $ax + by = 1$, then $a^2(x^2) + ab(2xy) + b^2(y^2) = 1$. So any common divisor of a^2, ab and b^2 must divide 1. This shows that the result is true.

(b) Let $d = (a + b, a^2 - ab + b^2)$. Then, d divides

$$(a + b) \cdot (a + b) - (a^2 - ab + b^2) = 3ab.$$

A common divisor of a and $a + b$ is also a divisor of b. Since $(a, b) = 1$, this shows that $(a + b, a) = 1$. Hence $(d, a) = 1$. Similarly, $(d, b) = 1$. Since $d|3ab$, this implies that $d|3$, and hence $d = 1$ or 3. Thus, (b) is true.

(c) If $a = 2$ and $b = 5$, then $b = a^2 + 1$ and $a^4 + 1 = 17$. So (c) is false.

(d) We can see that 25 cannot be expressed as $p + a^2$, where $p = 1$ or prime. The possible values of a are 0, 1, 2, 3, 4 and 5. Correspondingly, $25 - a^2 = 25, 24, 21, 16$ and 0, none of which can be p. Thus, (d) is false.

(e) If p is prime and $p > 3$, then $p = 6k \pm 1$, and so $p^2 + 2 = 36k^2 \pm 12k + 1 + 2$ is divisible by 3 and is greater than 3. Hence (e) is true.

10. If $\{p_n\}$ is a sequence of distinct primes, prove that

$$\frac{1}{p_1} + \frac{1}{p_2} + \cdots + \frac{1}{p_n}$$

is never an integer.

Solution If the sum is s, then

$$p_1 s - 1 = p_1 \left(\frac{1}{p_2} + \cdots + \frac{1}{p_n} \right) = p_1 \frac{a}{b},$$

where b is not divisible by p_1. Thus, $bp_1 s - b = p_1 a$. If s is an integer, this implies that $p_1 | b$.

11. Let $R_n = 11 \cdots 1$ in decimal notation, where the digit 1 is repeated n times. Prove that

(a) $R_n | R_m \iff n | m$,

(b) if $(n, m) = 1$, then $(R_n, R_m) = 1$.

(c) Find all n such that R_n is a perfect square.

Solution (a) Note that $R_n = 10^{n-1} + \cdots + 10 + 1 = \dfrac{10^n - 1}{10 - 1}$. So $\dfrac{R_m}{R_n} = \dfrac{10^m - 1}{10^n - 1}$.

Let $m = kn + r$, where $0 \leq r < n$, and $x = 10^n$. Then,

$$10^m - 1 = x^k \cdot 10^r - 1 = (x^k - 1)10^r + (10^r - 1).$$

Since $x^k - 1$ is divisible by $x - 1$, we see that $10^m - 1$ is divisible by $10^n - 1 = x - 1$ if and only if $10^r - 1$ is divisible by $x - 1$. But $0 \leq 10^r - 1 < x - 1$ as $0 \leq r < n$. Hence, $10^m - 1$ is divisible by $10^n - 1$ if and only if $10^r - 1 = 0$, that is, $r = 0$. This proves (a).

(b) There exist positive integers x and y such that $mx - ny = 1$. Now, $10^{mx} - 1 = a^x - 1$, where $a = 10^m$. This is divisible by $a - 1 = 9R_m$. Similarly, $10^{ny} - 1$ is divisible by $9R_n$. We also have $10^{mx} = 10 \cdot 10^{ny}$, and so

$$10^{mx} - 1 = 10\,(10^{ny} - 1) + (10 - 1).$$

Thus, $u(9R_m) = 10v(9R_n) + 9$ with integers u, v. This gives

$$uR_m - (10v)R_n = 1,$$

showing that $(R_m, R_n) = 1$.

(c) If $n = 1$, then $R_n = 1$ is a perfect square. Let $n > 1$. Then, $R_n - 11$ is divisible by 100, and hence by 4. So $R_n \equiv 3 \pmod 4$. But the square of an odd integer is of the form $4k + 1$, and the square of an even integer is divisible by 4. Thus, no perfect square can be of the form $4k + 3$. Hence, R_n cannot be a perfect square if $n > 1$.

12. Find the least integer $n > 1$ such that the number $9997n$ contains only odd digits in decimal notation.

Solution Clearly, n must be odd, $n \geq 3$. Suppose that $n \leq 3333$. Let

$$n = a_0 10^3 + a_1 10^2 + a_2 10 + a_3, \quad 0 \leq a_0 \leq 3, \quad 0 \leq a_i \leq 9.$$

Since $9 \leq 3n \leq 9999$, we have $1 \leq 10000 - 3n \leq 9991$. Let

$$10000 - 3n = b_0 10^3 + b_1 10^2 + b_2 10 + b_3, \quad 0 \leq b_i \leq 9.$$

Then,

$$\begin{aligned} 9997n = (10000 - 3)n &= a_0 10^7 + a_1 10^6 + a_2 10^5 + (a_3 - 1)10^4 \\ &\quad + b_0 10^3 + b_1 10^2 + b_2 10 + b_3. \end{aligned}$$

Since n is odd, a_3 is odd. If $a_3 = 1$, then one of a_0, a_1, a_2 is non-zero. Hence, $9997n$ contains an even digit $a_3 - 1$ if $n \leq 3333$. If $n = 3335$, then $9997n = 3333\,9995$. Therefore, $n = 3335$ is the least value.

13. Find the least positive integer n such that the product of 9999 and n contains only the digit 1 in decimal notation.

Solution Let $R_k = 11 \cdots 1$, where the digit 1 is repeated k times and $N = 9999$. Suppose that $Nn = R_k$ for some k. Since $R_k \equiv k \pmod 9$, we first observe that k is divisible by 9. Now, $R_k = (10^k -$

$1)/9$, and $N+1 = 10^4$. Suppose that $k = 4r + s$, where $s = 0, 1, 2$ or 3. Then,

$$9Nn = 10^{4r+s} - 1 = (10^4)^r \cdot 10^s - 1 \equiv 10^s - 1 \pmod{N},$$

and hence, $10^s - 1 \equiv 0 \pmod{N}$. If $s = 1, 2$ or 3, then $0 < 10^s - 1 < N$. This shows that $s = 0$, that is, k is divisible by 4. Thus, the least possible value of k is 36, and the least value of n is

$$\frac{R_{36}}{N} = \frac{10^{36} - 1}{9(10^4 - 1)}.$$

14. Prove that the sequence

$$(n+1)! - 2, (n+1)! - 3, \ldots, (n+1)! - (n+1)$$

produces n consecutive composite numbers for $n > 2$. Prove also that each term in this sequence is divisible by a prime which does not divide any other term in it.

Solution Let $x_k = (n+1)! - k$ for $2 \le k \le n+1$. Since $(n+1)!$ is divisible by k for $k \le n+1$, we see that x_k is also divisible by k. But $x_k > k$ since $n > 2$, and hence x_k is composite. This proves the first statement.

Let $a_k = \dfrac{x_k}{k} = \dfrac{(n+1)!}{k} - 1$. Then, $a_k > 2$ since $n > 2$. Let p_k be a prime divisor of a_k. Then $p_k | x_k$. Suppose that p_k divides x_r also, for some $r \ne k$. Then p_k divides $x_k - x_r = r - k$. Hence $p_k \le |r - k| < n + 1$, and $p_k | (n+1)! = x_r + r$. Thus, $p_k | r$. Since r divides $\dfrac{(n+1)!}{k} = a_k + 1$, this implies that $p_k | (a_k + 1)$. This is not possible since $p_k | a_k$. It follows that p_k does not divide any x_r other than x_k.

15. Prove that for any positive integers n and k, there exists a set of n consecutive integers such that each member of this set is divisible by k distinct primes, none of which divides any other member of the set.

Solution If $n = 1$, the result is trivial. We shall prove the result by induction on k. Let $k = 1$. If $n = 2$, we have two numbers 8, 9 with distinct prime divisors. If $n > 2$, we have n consecutive composite numbers with the desired property, by the previous problem.

Assume that, for some k, there exist n consecutive composite numbers

$$N, N-1, N-2, \ldots, N-n+1 \tag{1}$$

such that the ith term $N - i + 1$ is divisible by k distinct primes $p_{i1}, p_{i2}, \ldots$, none of which divides any other term in it. Consider a new sequence

$$N! + N, N! + N - 1, N! + N - 2, \ldots, N! + N - n + 1 \qquad (2)$$

of n consecutive composite numbers. Let $P_i = N!/(N - i + 1)$, which is the product of all numbers in the set $S = \{2, 3, \ldots, N\}$ other than $N - i + 1$. Since $N - i + 1$ is not a prime, we see that if p is any prime in the set S, then p divides P_i. Let q_i be a prime divisor of $P_i + 1$. Then, q_i divides $(N - i + 1)(P_i + 1)$ = the ith term in (2). Suppose that q_i divides the jth term $N! + N - j + 1$ also, and $j \neq i$. Then $q_i | (j - i)$, and so $q_i \leq |j - i| \leq n - 1 < N$, in which case, q_i must divide P_i by what has been observed above. Since this is not possible, q_i divides only the ith term in (2). Since $p_{ij} < N - i + 1$, we see that p_{ij} divides P_i, and hence, $p_{ij} \neq q_i$ for any j. Further, p_{ij} divides $N!$ for all i, j. Thus, we see that the ith term in (2) has $k + 1$ distinct prime divisors $q_i, p_{i1}, p_{i2}, \ldots$, none of which divides any other term in it. This completes the proof by induction.

16. Consider an arithmetic progression $p, p + d, \ldots, p + (n - 1)d$. Prove that if all the n terms are prime numbers, and $n > 2$, then the common difference d is divisible by every prime $q < n$.

Solution Suppose that all the terms are prime, and let q be a prime $< n$. Assume that d is not divisible by q. Then the congruence $dx + p \equiv 0 \pmod{q}$ has a solution in $\{0, 1, \ldots, q - 1\}$; that is, $p + dk$ is divisible by q for some k with $0 \leq k \leq q - 1$. Then $p + kd$ is one of the n terms, and hence prime. So $p + kd = q$. Since $p + kd \geq p$, and $q \leq n - 1$, we get $p \leq n - 1$. In that case, $p + pd$ is one of the n terms. Since $p + pd$ is composite, we get a contradiction. This gives the result.

17. Find an arithmetic progression of 6 terms, each of which is a prime.

Solution In view of the above result, we have to choose a common divisor d that is divisible by $2 \cdot 3 \cdot 5$, and a prime $p > 6$ as the first term. With $d = 30, p = 7$, we get the progression $7, 37, 67, 97, 127, 157, \ldots$, giving 6 terms that are primes.

With $p = 199$ and $d = 210$, we get an arithmetic progression of 10 primes: 199, 409, 619, 829, 1039, 1249, 1459, 1669, 1879, 2089.

With $p = 166601$ and $d = 11550$, we have an arithmetic progression of 12 primes; and with $p = 4943$ and $d = 60060$, we have one of 13 primes.

18. Let k be a positive integer. Prove that the arithmetic progression

$$a, a + d, a + 2d, a + 3d, \ldots$$

of positive integers, where $(a, d) = 1$, contains k consecutive terms that are composite.

Solution Let $t_r = a + rd$. Then, $t_{n+r} = nd + t_r$. So if n is divisible by t_r, then t_{n+r} is divisible by t_r, and is, therefore, composite if $t_r > 1$. Hence, if we take $n = t_1 t_2 \cdots t_k$, then $t_{n+1}, t_{n+2}, \ldots, t_{n+k}$ are composite.

19. Given any positive integer k, prove that there are k consecutive integers each divisible by a square greater than 1.

Solution Let $p_1, p_2, \ldots, p_k$ be distinct primes. Consider the following congruences:

$$x \equiv -r + 1 \pmod{p_r^2} \quad \text{for } 1 \leq r \leq k.$$

By the Chinese remainder theorem, this simultaneous system has a solution. If $x = m$ is a solution, then $m + r - 1$ is divisible by p_r^2 for $1 \leq r \leq k$. So the consecutive integers $m, m + 1, \ldots, m + k - 1$ have the desired property.

20. Use the Chinese remainder theorem to find five consecutive terms in the arithmetic progression $1, 6, 11, 16, \ldots$, each of which is divisible by a prime which does not divide another one.

Solution The terms are $5k + 1$, $k = 0, 1, 2, \ldots$. We can choose primes (other than 5) conveniently. Let us choose n such that

$$5n + 1 \equiv 0 \pmod 2, \quad 5n + 6 \equiv 0 \pmod 3,$$
$$5n + 11 \equiv 0 \pmod 7, \quad 5n + 16 \equiv 0 \pmod{11},$$
$$5n + 21 \equiv 0 \pmod{13}.$$

First find the 'reciprocals' of 5 modulo these primes. We have $5 \cdot 1 \equiv 1 \pmod 2$, $5 \cdot 2 \equiv 1 \pmod 3$, $5 \cdot 3 \equiv 1 \pmod 7$, $5 \cdot 9 \equiv 1 \pmod{11}$ and $5 \cdot 8 \equiv 1 \pmod{13}$. So the 'reciprocals' are 1, 2, 3, 9 and 8 respectively. Hence the desired congruences can be written as

$$n \equiv -1 \pmod 2, \quad n \equiv -6 \cdot 2 \equiv 0 \pmod 3,$$
$$n \equiv -11 \cdot 3 \equiv 2 \pmod 7, \quad n \equiv -16 \cdot 9 \equiv 10 \pmod{11},$$
$$n \equiv -21 \cdot 8 \equiv 1 \pmod{13}.$$

The least positive solution is $n = 2991$, and the corresponding consecutive terms are 14956, 14961, 14966, 14971, 14976.

21. Determine all primes that can divide two successive integers of the form $n^2 + 3$.

Solution Suppose that n^2+3 and $(n+1)^2+3$ are divisible by a prime p. Then their difference is divisible by p, that is, $2n + 1 = pk$ for some k. So $4n^2 = (pk - 1)^2 \equiv 1 \pmod{p}$. Since $4(n^2 + 3) \equiv 0 \pmod{p}$, we get $1 + 12 \equiv 0 \pmod{p}$. This shows that $p = 13$. Now, 13 has the desired property, since $6^2 + 3$ and $7^2 + 3$ are divisible by 13.

22. The nth triangular number t_n is the sum of the first n natural numbers. Determine all twin primes $p, p+2$ whose arithmetic mean is a triangular number.

Solution Let $t_n = \frac{1}{2}\{p + (p+2)\} = p + 1$. Then,

$$p = \frac{1}{2}n(n+1) - 1 = \frac{1}{2}(n+2)(n-1).$$

If $n = 2k$ is even, then this becomes $p = (k+1)(2k-1)$, giving $2k - 1 = 1$, $k = 1$ and $p = 2$. If $n = 2k + 1$, then $p = (2k+3)k$, which gives $k = 1$ and $p = 5$. Thus, we get only one pair (5,7) of primes whose mean is a triangular number $6 = t_3$.

23. Determine all twin primes p and $q = p + 2$ such that $pq - 2$ is also prime.

Solution If $p = 3$, then $q = p+2 = 5$ and $pq-2 = 13$ are primes. Let $p > 3$. If $p \equiv 1 \pmod{3}$, then $p + 2$ is not prime. If $p \equiv 2 \pmod{3}$, then $q \equiv 1 \pmod{3}$, and so $pq - 2 \equiv 2 - 2 \pmod{3}$, $pq - 2$ is not prime. Thus, $p = 3$ is the only prime with the desired property.

24. (a) Using Bertrand's theorem, or otherwise, show that if $n > 1$, then $n!$ is never a perfect square.

(b) Find all positive integers n such that $n! + (n+1)! + (n+2)!$ is a perfect square.

Solution (a) If $n = 2$, then $n! = 2$ is not a perfect square. Let $n > 2$. By Bertrand's theorem, there is a prime p with $n/2 < p < n$. So $n!$ is divisible by p, but not by p^2. Hence, $n!$ cannot be a perfect square.

(b) We have $n + (n+1) + (n+2) = n!\{1 + (n+1) +(n+1)(n+2)\} = n!(n+2)^2$. This is a perfect square if and only if $n!$ is a perfect square, that is, $n = 1$.

Alternate solution of (a) *without assuming Bertrand's theorem*:[1] Assume that $n!$ is a perfect square for some $n > 1$. Let $k = [n/2]$ and

$$N = \binom{n}{k}, \quad N_1 = N\frac{(n-k)!}{k!} = \frac{n!}{(k!)^2}.$$

If n is even, then $n = 2k$ and $N_1 = N$; if n is odd, then $n = 2k+1$ and $N_1 = N(k+1)$. Thus, N_1 is a positive integer and must be a perfect square.

Now, $N = \prod_{p|N} p^{u_p}$, where

$$u_p = \sum_{i\geq 1}\left\{\left[\frac{n}{p^i}\right] - \left[\frac{k}{p^i}\right] - \left[\frac{n-k}{p^i}\right]\right\}.$$

Since $[\alpha+\beta] - [\alpha] - [\beta] = 0$ or 1, we see that each term in this sum is either 0 or 1. If $p^r \leq n < p^{r+1}$, then

$$0 < \frac{k}{p^i} \leq \frac{n-k}{p^i} < \frac{n}{p^i} < 1 \quad \text{for } i \geq r+1.$$

Hence,

$$\left[\frac{n}{p^i}\right] = \left[\frac{k}{p^i}\right] = \left[\frac{n-k}{p^i}\right] = 0 \quad \text{for } i \geq r+1,$$

and $u_p \leq r$, where r is such that

$$r \log p \leq \log n < (r+1)\log p.$$

We shall prove that $p \leq \sqrt{n}$ for all prime divisors p of N. Suppose that $u_p = 1$. Then, p divides N and so $p|N_1$ since $N|N_1$. Since N_1 is a perfect square, we see that p^2 divides N_1. But p^2 does not divide N if $u_p = 1$. This implies that $N_1 \neq N$, n is odd and that $p|k+1$. Let $k+1 = ap$. If $r = 1$, then $n < p^2$, and

$$u_p = \left[\frac{n}{p}\right] - \left[\frac{k}{p}\right] - \left[\frac{k+1}{p}\right] = \left[\frac{2ap-1}{p}\right] - \left[\frac{ap-1}{p}\right] - a = 0.$$

Consequently, $u_p = 1$ implies that $r \geq 2$, and hence $p^2 \leq p^r \leq n$, as desired. Next suppose that $u_p \geq 2$. Then, $2 \leq u_p \leq r$, and so

$$2\log p \leq u_p \log p \leq r\log p \leq \log n;$$

[1] This alternate solution is given here following a query whether such a strong property like Bertrand's theorem is needed for the solution.

hence, $p^2 \leq n$ again.

The number of primes p with $p \leq \sqrt{n}$ is less than $\sqrt{n}$. Since $u_p \leq r$, we get

$$\log N = \sum_{p|N} u_p \log p \leq \sum_{p|N} r \log p \leq (\log n)\sqrt{n}. \qquad (*)$$

Note that if $n = 2k$, then $N = \binom{n}{k}$ is the largest term in the binomial expansion of $(1+1)^n$. If $n = 2k+1$, then $N = \binom{n}{k}$ is one of the two largest terms in this expansion. Hence,

$$2^n = 2 + \binom{n}{1} + \binom{n}{2} + \cdots + \binom{n}{n-1} \leq n\binom{n}{k} = nN,$$

$$n \log 2 \leq \log n + \log N \leq \log n + (\log n)\sqrt{n},$$

using (*). Divide by log 2, and write $x = \sqrt{n}$ and $y = \dfrac{\log n}{\log 2}$ so that $\log n = y \log 2$. We get

$$x^2 \leq y + yx. \qquad (**)$$

Since $2^y = n$, we get $x = \sqrt{n} = 2^{y/2}$. Let $m = [y/2]$. Suppose that $n \geq 256$ so that $y \geq 8$. Since $2^s \geq s+1$ for all $s \geq 1$, and $m = [y/2] \geq 4$, we see that

$$\begin{aligned} x = 2^{y/2} \geq 2^m = 4(2^{m-2}) \geq 4(m-1) \geq 2m+4 \\ = 2(m+1) + 2 > y + 2. \end{aligned}$$

It follows that

$$x^2 > yx + 2x > yx + 2(y+2),$$

which contradicts (**).

Assuming that $n!$ is a perfect square, we have obtained a contradiction if $n \geq 256$. Obviously, $n!$ is not a perfect square for $n = 2, 3$. If p is a prime with $n/2 < p < n$, then $n!$ is divisible by p but not by p^2; hence, $n!$ cannot be a perfect square if such a prime exists. Consider the primes p = 3, 5, 7, 13, 23, 43, 83, 163, each of which is less than twice the preceding. We can see that for any n with $1 < n < 256$, there exists such a prime. This proves that $n!$ cannot be a perfect square for any $n > 1$.

25. For $n \equiv \pm 4 \pmod 9$, show that the equation $n = a^3 + b^3 + c^3$ has no integer solutions.

Solution Let $n = a^3 + b^3 + c^3$. If $x = 3k + r$, then

$$x^3 = 27k^3 + 27k^2r + 9kr^2 + r^3 \equiv r^3 \pmod 9.$$

Hence $x^3 \equiv 0, 1$ or 8 medal 9 according as $x \equiv 0, 1$ or $2 \pmod 3$. If each of a, b, c is divisible by 3, then $n \equiv 0 + 0 + 0 \pmod 9$. If exactly two of them are divisible by 3, then $n \equiv 1$ or $8 \pmod 9$. If exactly one of them is divisible by 3, then $n \equiv 1 + 1, 1 + 8$ or $8 + 8 \pmod 9$. Finally, if none of them is divisible by 3, then $n \equiv 3, 10, 17$ or $24 \pmod 9$. In none of these cases do we have $n \equiv \pm 4 \pmod 9$.

26. Show that each term of the sequence

$$16, 1156, 111556, 11115556, 1111155556, \ldots$$

is a perfect square.

Solution Let t_n be the nth term and $R_k = 111 \cdots 1$ in decimal notation, where the digit 1 is repeated k times. Then, $t_1 - 1 = 15$, $t_2 - 1 = 1155$. In general, $t_n - 1$ consists of $2n$ digits of which the first n are 1, and the remaining are 5. Hence,

$$t_n - 1 = R_n \cdot 10^n + R_n \cdot 5 = R_n(10^n + 5).$$

Since $R_k = (10^k - 1)/9$, we get

$$9t_n - 9 = (10^n - 1)(10^n + 5) = 10^{2n} + 4 \cdot 10^n - 5.$$

It follows that $9t_n = (10^n+2)^2 = (9R_n+3)^2$. Hence, $t_n = (3R_n+1)^2$.

27. Find all primes of the form $p^2 + 2^p$, where p is prime.

Solution If $p = 2$, then $p^2 + 2^p = 8$, and if $p = 3$, then $p^2 + 2^p = 17$. Let p be a prime > 3. Then, $p^2 \equiv 1 \pmod 3$. Since p is odd, we have $p = 2k + 1$ and $2^p = 4^k \cdot 2 \equiv 2 \pmod 3$. Hence, $p^2 + 2^p \equiv 1 + 2 \pmod 3$. Thus, if p is a prime > 3, then $p^2 + 2^p$ is divisible by 3, and hence composite. So $p = 3$ is the only prime such that $p^2 + 2^p$ is prime.

28. The primes $37, 67, 73, 79, \ldots$ are of the form $p = 36ab + 6a - 6b + 1$, with $a \geq 1$, $b \geq 1$. Show that no pair of twin primes can contain a prime of this form.

Solution Let $p = 36ab + 6a - 6b + 1 = (6a - 1)(6b + 1) + 2$. Then, $p - 2$ has divisors $6a - 1$ and $6b + 1$, both of which are > 1 if $a \geq 1$, $b \geq 1$. On the other hand, $p + 2 = 3(12ab + 2a - 2b + 1)$ is composite since $12ab + 2a - 2b + 1 > 2a + 1 > 1$. Thus, we see that $p - 2$ and $p + 2$ are composite numbers. This proves the result.

29. Let $p_1, p_2, \ldots, p_n$ be the first n primes in the natural order. Show, for $n \geq 2$, that there are at least two new primes in the interval $p_n < x \leq p_1 p_2 \cdots p_n + 1$.

    Solution Let $P_n = p_1 p_2 \cdots p_n + 1$. Then, $P_n - 2 \geq 2p_n - 2 > p_n$ for $n \geq 2$. So, P_n and $P_n - 2$ are odd numbers greater than 3, and have prime divisors p and q, say. Since p and q are odd, and $P_n - (P_n - 2) = 2$, we see that $p \neq q$. Thus, we get two distinct primes p and q, in the interval specified by $p_n < x \leq P_n$.

30. (a) Find the last two digits in the decimal representation of the number 9^{9^9}.

    (b) Find the last three digits in the decimal representation of the number $9^{9^{9^9}}$.

    Solution (a) Using binomial theorem we get,

$$N = 9^9 = (10 - 1)^9 \equiv (-1)^9 \equiv 9 \pmod{10},$$

$$N_1 = 9^{9^9} = (10 - 1)^N \equiv (-1)^N + N(-1)^{N-1} \cdot 10 \pmod{100}.$$

Thus, $N = 10k + 9$, $(-1)^N = -1$ and $10N \equiv 90 \pmod{100}$. Hence, $N_1 \equiv -1 + 90 \equiv 89 \pmod{100}$. The last two digits in 9^{9^9} are 89.

(b) By (a), $N_1 = 9^{9^9} \equiv 89 \pmod{100}$. Let $N_1 = 100k + 89$. Then, by binomial theorem,

$$9^{N_1} \equiv -1 + N_1 \cdot 10 - \binom{N_1}{2} \cdot 10^2 \pmod{1000}.$$

Now, $10N_1 = 10(100k + 89) \equiv 890 \pmod{1000}$, and

$$100\binom{N_1}{2} = 100(100k + 89)(50k + 44) \equiv 391600 \pmod{1000}.$$

Hence, $9^{9^{9^9}} \equiv -1 + 890 - 600 \equiv 289 \pmod{1000}$. The last three digits in $9^{9^{9^9}}$ are 289.

Set II

1. Given integers x and y, let $m = ax + by$ and $n = cx + dy$, where a, b, c, d are integers such that $ad - bc = \pm 1$. Prove that $\gcd(m, n) = \gcd(x, y)$.

   Solution Let $u = \gcd(m, n)$ and $v = \gcd(x, y)$. Since v divides x and y, it divides $ax + by = m$ and $cx + dy = n$, and hence v divides $\gcd(m, n) = u$. Now,

   $$dm - bn = (ad - bc)x = \pm x, \quad cm - an = (bc - ad)y = \pm y.$$

   As before, we can see that u divides x and y, and hence, v. It follows that $u = v$.

2. If a, m, n are positive integers and $m \neq n$, find $\gcd(a^{2^m} + 1, a^{2^n} + 1)$.

   Solution Let $x_n = a^{2^n} + 1$ and $m > n$, that is, $m \geq n + 1$. Then, $2^m = 2^n(2b)$ for some integer b, and hence,

   $$x_m - 1 = a^{2^m} = \left(a^{2^n}\right)^{2b} = (x_n - 1)^{2b}.$$

   Expanding by binomial theorem we see that $x_m - 1$ is equal to $1+$ a multiple of x_n. Thus, $x_m - 2 = Nx_n$ for some integer N. Hence, $\gcd(x_m, x_n)$ divides 2. If a is odd, then x_n is even and if a is even, then x_n is odd, for all n. It follows that if a is odd, then the gcd is 2, and if a is even, then it is 1.

3. Prove that every integer $n \geq 12$ is the sum of two composite numbers.

   Solution Consider the set $S = \{2, 3, \ldots, n - 2\}$ of $n - 3$ numbers. Suppose that k of them, say $c_1, c_2, \ldots, c_k$, are composite, and the remaining $n - 3 - k$ are primes.

   Assume that the k numbers $n - c_1, n - c_2, \ldots, n - c_k$ are all primes. Since these are distinct and the total number of primes in S is $n - 3 - k$, we must have $n - 3 - k \geq k$, that is, $2k \leq n - 3$. If $m = [n/2]$, then the $m - 2$ numbers in $T = \{2 \cdot 2, 2 \cdot 3, \ldots, 2(m - 1)\}$ are composite and are in S. If $n \geq 17$, then 9 and 15 are composite numbers in S but not in T. Hence, if $n \geq 17$ then S contains at least m composite numbers, that is, $k \geq m$. Thus,

   $$m \leq k \leq \frac{n}{2} - \frac{3}{2} < m + 1 - \frac{3}{2},$$

a contradiction. This shows that if $n \geq 17$, then at least one of the k numbers $n - c_i$ is a composite, and $n = c_i + (n - c_i)$.

Finally, $12 = 6 + 6$, $13 = 9 + 4$, $14 = 10 + 4$, $15 = 9 + 6$ and $16 = 10 + 6$. This completes the proof.

4. Let a, b, x, y be positive integers such that $\gcd(a, b) = 1$ and $x^a = y^b$. Prove that $x = n^b$ and $y = n^a$ for some n.

Solution If $(a, b) = 1$, then there exist integers u and v such that $au + bv = 1$. Let $n = x^v y^u$. Then, n is rational, and

$$n^b = x^{bv} y^{bu} = x^{bv}(x^a)^u = x^{au+bv} = x.$$

Thus, n is a positive rational number such that $n^b = x$, an integer. Hence, n is an integer. Further, $n^{ab} = x^a = y^b$, and hence, $n^a = y$.

5. Let a, b, n be positive integers such that $\gcd(a, b) = 1$ and $n > ab$. Prove that there exist positive integers x and y such that $n = ax + by$. Prove also that this is not true if $n = ab$.

Solution There are integers u and v such that $au + bv = 1$. So

$$n = n(au + bv) = a(nu + kb) + b(nv - ka), \quad \text{for all integers } k.$$

If $x = nu + kb$ and $y = nv - ka$, then, x and y are integers such that $ax + by = n$. We shall choose the integer k such that x and y are positive. Now,

$$\frac{nv}{a} + \frac{nu}{b} = \frac{n}{ab}(au + bv) = \frac{n}{ab} > 1,$$

since $n > ab$. Let $k = [-nu/b] + 1$. Then, $k > -nu/b$ and $k \leq -nu/b + 1 < nv/a$. Hence, $x > 0$ and $y > 0$. This proves the first part.

For the second part, suppose that $ab = ax + by$. Since $\gcd(a, b) = 1$, this implies that $a|y$ and $b|x$. Hence, $y \geq a$ and $x \geq b$ since a, b, x, y are positive integers. In that case, $ax + by \geq 2ab$, contradicting our assumption.

6. If $n > 1$ is an integer not of the form $6k + 3$, prove that $n^2 + 2^n$ is composite.

Solution If n is even, then $n^2 + 2^n$ is even. If n is odd and not of the form $6k + 3$, then it is of the form $6k \pm 1$ and so $n^2 \equiv 1 \pmod 3$. If n

is odd and $n = 2r + 1$, then $2^n = 4^r \cdot 2 \equiv 1 \cdot 2 \pmod 3$. Hence if n is odd, then $n^2 + 2^n \equiv 1 + 2 \pmod 3$. Thus, $n^2 + 2^n$ is divisible by 2 or 3 if n is not of the form $6k + 3$. Note that $n^2 + 2^n$ is prime when $n = 3$ and 9.

7. If p is any prime other than 2 or 5, prove that

   (a) p divides infinitely many terms in the sequence $9, 99, 999, \ldots$.

   (b) p divides infinitely many terms in the sequence $1, 11, 111, \ldots$.

   Solution If a_n is the nth term in (a) and b_n is the nth term in (b), then $a_n = 10^n - 1$ and $b_n = a_n/9$. Since $(10, p) = 1$, we have $10^{p-1} \equiv 1 \pmod p$, and hence $10^{k(p-1)} \equiv 1 \pmod p$. Thus, $a_n \equiv 0 \pmod p$ whenever $n = k(p-1)$, $k \geq 1$. This proves (a).

   Further, $9b_n = a_n \equiv 0 \pmod p$ for $n = k(p-1)$. Hence, if $p \neq 3$, then $b_n \equiv 0 \pmod p$ for these values of n. If $p = 3$, then $b_{3k} \equiv 0 \pmod p$. This proves (b).

8. For any prime p, if $a^p \equiv b^p \pmod p$, prove that $a^p \equiv b^p \pmod{p^2}$.

   Solution We have $a^p - b^p = (a - b)\left(a^{p-1} + a^{p-2}b + \cdots + ab^{p-2} + b^{p-1}\right)$. Also, $a^p \equiv a$ and $b^p \equiv b \pmod p$. So $a \equiv b \pmod p$. Thus the first factor, namely $a-b$, is divisible by p. Moreover, $a^k b^{p-1-k} \equiv a^{p-1} \pmod p$ for $k = 0, 1, 2, \ldots$. So the second factor is congruent to pa^{p-1} modulo p. Thus, both factors are divisible by p, and hence the product is divisible by p^2.

9. Given an integer n, prove that there is an integer m divisible by n such that the decimal representation of m contains only the digits 0 and 1. Prove also that the same is true for digits 0 and 2, or 0 and 3, ..., or 0 and 9, but for no other pair of digits.

   Solution Let $R_k = 11 \cdots 1$, where 1 is repeated k times. Consider the n numbers $R_1, R_2, \ldots, R_n$. Either these are incongruent modulo n, or else, some two of them are congruent modulo n. In the first case, some R_i is congruent to 0 modulo n; in the other case, $R_i \equiv R_j \pmod n$, where $1 \leq i < j \leq n$, and so $R_j - R_i \equiv 0 \pmod n$. Note that R_i consists of the digit 1 only, and $R_j - R_i = 1 \cdots 10 \cdots$ consists of 1 and 0 only. This gives the first statement.

   Instead of $R_1, \ldots, R_n$, consider the numbers $aR_1, \ldots, aR_n$, where $2 \leq a \leq 9$. We see, exactly as above, that n divides some m whose only digits are 0 and a. On the other hand, any multiple of 10 has 0 as its end digit. So, if $n = 10$, then every multiple of n has zero as one of its digits.

10. For $n \geq 1$, show that

$$(-13)^{n+1} \equiv (-13)^n + (-13)^{n-1} \pmod{181}.$$

Determine all integers a such that the congruence $a^{n+1} \equiv a^n + a^{n-1} \pmod{181}$ holds for all positive integers n.

Solution We have

$$a^2 - a - 1 \equiv a^2 - a - 182 \pmod{181},$$

and $a^2 - a - 182 = (a-14)(a+13)$. Thus, $a^2 - a - 1 \equiv 0 \pmod{181}$ if and only if $(a - 14)(a + 13) \equiv 0 \pmod{181}$, that is, $a + 13 \equiv 0 \pmod{181}$ or $a - 14 \equiv 0 \pmod{181}$. If $a^2 - a - 1 \equiv 0 \pmod{181}$, then

$$a^{n+1} - a^n - a^{n-1} \equiv 0 \pmod{181}$$

for all positive integers n. It follows that $a^{n+1} \equiv a^n + a^{n-1} \pmod{181}$ for all $n \geq 1$ if and only if $a \equiv -13 \pmod{181}$ or $a \equiv 14 \pmod{181}$.

11. Prove that if a is an odd integer, then $a^{2^n} \equiv 1 \pmod{2^{n+2}}$ for all positive integers n.

Solution If $a = 2k + 1$, then $a^2 = 4k(k + 1) + 1 \equiv 1 \pmod{8}$ since $k(k + 1)$ is always even. Hence, the assertion is true when $n = 1$.

Assume that it is true for some $n \geq 1$. Then, $a^{2^n} = 2^{n+2}m + 1$ for some integer m. This implies that

$$\begin{aligned} a^{2^{n+1}} &= \left(a^{2^n}\right)^2 = \left(2^{n+2}m + 1\right)^2 = \left(2^{2n+4}m^2 + 2(2^{n+2}m) + 1\right) \\ &\equiv 1 \pmod{2^{n+3}}. \end{aligned}$$

This proves the result by induction.

12. If p and q are distinct primes, prove that $p^{q-1} + q^{p-1} \equiv 1 \pmod{pq}$.

Solution We have $p^{q-1} \equiv 1 \pmod{q}$, by Fermat's theorem, since p is not divisible by q, and $q^{p-1} \equiv 0 \pmod{q}$. Thus, $p^{q-1} + q^{p-1} \equiv 1 + 0 \pmod{q}$. Similarly, $q^{p-1} + p^{q-1} \equiv 1 \pmod{p}$ since q is not divisible by p. Thus, $p^{q-1} + q^{p-1} - 1$ is divisible by p and q, and hence by pq.

13. Show that there exist no primes p and q that satisfy the condition $p^2 = 10^q - 999$.

Solution If $q = 2$, then $10^q - 999 < 0$. If $p = 7$, then $p^2 + 999 = 1048 \neq 10^q$ for any q. Let p be a prime other than 7 and q be an odd prime. Then, $p^2 \equiv 1, 2 \text{ or } 4 \pmod 7$. Since $999 \equiv 5 \pmod 7$, we see that

$$p^2 + 999 \equiv 6, 0 \text{ or } 2 \pmod 7.$$

Moreover, q is of the form $6k + 1$ or $6k + 5$. Since $10^6 \equiv 1 \pmod 7$, we see that if $q = 6k + 1$, then

$$10^q = (10^6)^k \cdot 10 \equiv 10 \equiv 3 \pmod 7,$$

and if $q = 6k + 5$, then

$$10^q = (10^6)^k \cdot 10^5 \equiv 10^5 \equiv 5 \pmod 7.$$

Thus, $p^2 + 999 \not\equiv 10^q \pmod 7$. This proves the result.

14. Find the last two digits in the decimal representation of $3^{1001} + 2^{1001}$.

Solution We are required to find the least non-negative integer r such that

$$3^{1001} + 2^{1001} \equiv r \pmod{100}.$$

Since $100 = 4 \times 25$, we have $\phi(100) = \phi(4)\phi(25) = 2 \times 20 = 40$. Hence, by Euler's theorem, $3^{40} \equiv 1 \pmod{100}$. It follows that $3^{1000} \equiv 1 \pmod{100}$, and $3^{1001} \equiv 3 \pmod{100}$.

Since 2 is not relatively prime to 100 we proceed as follows: We have $2^{20} \equiv 1 \pmod{25}$, and so $2^{1000} \equiv 1 \pmod{25}$. Thus,

$$2^{999}(2 \cdot 13) \equiv 13 \pmod{25}, \quad 2^{999} \equiv 13 \pmod{25}.$$

This implies that $2^{999} = 25m + 13$, and so $2^{1001} = 100m + 52$. Combining these two, we get $3^{1001} + 2^{1001} \equiv 3 + 52 \equiv 55 \pmod{100}$. So the last two digits are 55.

15. Let p be an odd prime, and $r_1, r_2, \ldots, r_{p-1}$ be the integers $1, 2, \ldots, p-1$ arranged in some order. Prove that at least two of the numbers

$$r_1,\ 2r_2,\ \ldots,\ (p-1)r_{p-1}$$

are congruent modulo p.

Solution Let $i \cdot r_i = s_i$, $1 \le i \le p-1$. Suppose that $s_1, \dots, s_{p-1}$ are incongruent modulo p. Since none of them is divisible by p, we can find t_i such that, for $1 \le i \le p-1$,

$$s_i \equiv t_i \pmod p,\ 0 < t_i < p.$$

Then the t_i are distinct elements in $\{1, 2, \dots, p-1\}$. Hence their product is $(p-1)\,!$. But their product is congruent modulo p to

$$s_1 \cdot s_2 \cdots s_{p-1} = (p-1)\,! \cdot r_1 \cdot r_2 \cdots r_{p-1}.$$

Since the product $r_1 \cdot r_2 \cdots r_{p-1}$ is also equal to $(p-1)\,!$, we get

$$(p-1)\,! \equiv ((p-1)\,!)^2 \pmod p.$$

Since $(p-1)\,! \equiv -1 \pmod p$, we get $-1 \equiv 1 \pmod p$, which is possible only when $p = 2$. This contradiction proves the result.

16. If n is a positive integer and x is real, prove that

$$[x] + \left[x + \frac{1}{n}\right] + \cdots + \left[x + \frac{n-1}{n}\right] = [nx].$$

Solution Let $[nx] = a$. Divide a by n. Let $a = mn + r$, where $0 \le r < n$. Then, $a + s \le nx + s < a + s + 1$. It follows that

$$x + \frac{s}{n} < m + \frac{r+s+1}{n} < m + 2 \quad \text{for } 0 \le r, s \le n-1,$$

and, further, $x + \frac{s}{n} < m + 1$ if $r + s + 1 \le n$. Thus we see that $[x + \frac{s}{n}] = m$ for $0 \le s \le n - r - 1$, and $[x + \frac{s}{n}] = m + 1$ for $n - r \le s \le n - 1$. Hence, the sum in the given relation is equal to $m(n-r) + (m+1)r = mn + r = a = [nx]$.

17. Determine the set of all real numbers α such that

$$\left[\sqrt{n+\alpha} + \tfrac{1}{2}\right] = \left[\sqrt{n} + \tfrac{1}{2}\right] \quad \text{for all positive integers } n.$$

Solution Let n be a positive integer and $k = \left[\sqrt{n} + \frac{1}{2}\right]$. Then,

$$k \le \sqrt{n} + \tfrac{1}{2} < k + 1, \quad k - \tfrac{1}{2} \le \sqrt{n} < k + \tfrac{1}{2}.$$

Hence,

$$k^2 - k + 1 \le n \le k^2 + k. \tag{1}$$

A real number α satisfies the given condition if and only if

$$[\sqrt{n+\alpha}+\tfrac{1}{2}]=k,$$

that is,

$$k^2-k+\tfrac{1}{4}\le n+\alpha<k^2+k+\tfrac{1}{4}. \tag{2}$$

This must be true when $n=k^2-k+1$ and also when $n=k^2+k$. The first value gives $\alpha\ge\frac{1}{4}-1$, and the other gives $\alpha<\frac{1}{4}$. Thus, $-\frac{3}{4}\le\alpha<\frac{1}{4}$. On the other hand, if $-\frac{3}{4}\le\alpha<\frac{1}{4}$ then, by (1),

$$k^2-k+1-\tfrac{3}{4}\le n+\alpha<k^2+k+\tfrac{1}{4},$$

which is the same as (2). Thus, we see that α satisfies the condition if and only if $-\frac{3}{4}\le\alpha<\frac{1}{4}$.

18. If $\gcd(m,n)=1$, prove that

$$\sum_{k=1}^{n-1}\left[\frac{mk}{n}\right]=\frac{(m-1)(n-1)}{2}.$$

Solution Let S be the set of all pairs (x,y) of integers with

$$1\le x\le m-1,\quad 1\le y\le n-1.$$

There are $(m-1)(n-1)$ elements in S. If $nx=my$, then $m|x$ and $n|y$ which implies that $x\ge m$ and $y\ge n$. This shows that $nx\ne my$ for any (x,y) in S. Let

$$A=\{(x,y)\in S: nx<my\},\quad B=\{(x,y)\in S: nx>my\}.$$

Then, $S=A\cup B$. Further,

$$nx<my\Longleftrightarrow n(m-x)>m(n-y).$$

If (x,y) belongs to S, then $(m-x,n-y)$ also belongs to S. Consequently,

$$(x,y)\in A\Longleftrightarrow(m-x,n-y)\in B.$$

This shows that A and B contain the same number elements. Hence, the number of elements in A is half the total number of elements in S, that is, $\frac{1}{2}(m-1)(n-1)$.

Moreover, $(x, y) \in A$ if and only if

$$1 \le x < \frac{my}{n}, \quad 1 \le y \le n-1.$$

Thus, for each fixed y there are $[my/n]$ values of x such that $(x, y) \in A$. It follows that the total number of elements in A is $\sum_{y=1}^{n-1} \left[\frac{my}{n}\right]$. This gives

$$\frac{1}{2}(m-1)(n-1) = \sum_{k=1}^{n-1} \left[\frac{mk}{n}\right].$$

19. Let θ be a real number such that $0 < \theta < 1$. Define

$$x_n = \begin{cases} 0 & \text{if } [n\theta] = [(n-1)\theta], \\ 1 & \text{otherwise.} \end{cases}$$

Prove that

$$\lim_{n\to\infty} \frac{x_1 + x_2 + \cdots + x_n}{n} = \theta.$$

Solution Let $a = [(n-1)\theta]$. Then, $a \le (n-1)\theta < a+1$. Adding θ,

$$a < a + \theta \le n\theta < a + 2,$$

since $0 < \theta < 1$. It follows that $[n\theta] = a$ or $a+1$. If $[n\theta] = a$, then $x_n = 0$, and if $[n\theta] = a+1$, then $[n\theta] \ne [(n-1)\theta]$, and so $x_n = 1$. This shows that $x_n = [n\theta] - [(n-1)\theta]$ for all n. Hence,

$$x_1 + x_2 + \cdots + x_n = [n\theta].$$

Since $n\theta - 1 < [n\theta] \le n\theta$, we get

$$\theta - \frac{1}{n} < \frac{x_1 + x_2 + \cdots + x_n}{n} \le \theta.$$

This proves the result.

20. Let n be an odd positive integer. If n factors into the product of two positive integers, $n = uv$, with $u > v$ and $u - v \le \sqrt[4]{64n}$, prove that the roots of the equation $x^2 - 2\left[\sqrt{n} + 1\right]x + n = 0$ are integers.

Solution Since $u > v$, we have $(u-v)^2 \le \sqrt{64n} = 8\sqrt{n}$. Hence,

$$(u+v)^2 = (u-v)^2 + 4uv \le 8\sqrt{n} + 4n < 4\left(\sqrt{n}+1\right)^2.$$

Further, since $u \neq v$, we also have

$$(u+v+2)^2 > (2\sqrt{uv}+2)^2 = 4\left(\sqrt{n}+1\right)^2.$$

It follows that
$$\frac{u+v}{2} < \sqrt{n}+1 < \frac{u+v}{2}+1.$$

But $(u+v)/2$ is an integer since u and v are odd. Thus, $[\sqrt{n}+1] = (u+v)/2$, and the equation reduces to $x^2-(u+v)x+uv=0$ whose roots are u and v.

21. Let α and β be positive irrational numbers such that $\dfrac{1}{\alpha}+\dfrac{1}{\beta}=1$. Prove that the two infinite sequences

$$[\alpha], [2\alpha], [3\alpha], \ldots \quad \text{and} \quad [\beta], [2\beta], [3\beta], \ldots$$

together contain each natural number exactly once.

Solution We shall first show that no positive integer occurs more than once in the two sequences taken together. If $[m\alpha]=[n\alpha]$ and $m>n$, then

$$m\alpha < [m\alpha]+1 = [n\alpha]+1 \leq n\alpha+1,$$

and hence, $1 > (m-n)\alpha \geq \alpha > 1$. Similarly, we get a contradiction if $[m\alpha] = [n\alpha]$ and $m < n$. Thus, no term in the first sequence is repeated. Similarly, we see that no term in the second sequence is repeated.

Next suppose that $[m\alpha] = k = [n\beta]$ for some m, n. Since α is irrational, we must have

$$k < m\alpha < k+1, \quad \frac{k}{m} < \alpha < \frac{k+1}{m}.$$

Similarly, since β is irrational, we have $k/n < \beta < (k+1)/n$. This shows that

$$\frac{m}{k+1}+\frac{n}{k+1} < \frac{1}{\alpha}+\frac{1}{\beta} < \frac{m}{k}+\frac{n}{k}.$$

Since $\dfrac{1}{\alpha}+\dfrac{1}{\beta}=1$, we get

$$\frac{m+n}{k+1} < 1 < \frac{m+n}{k}, \qquad k < m+n < k+1.$$

This is impossible. This proves the assertion.

Next, to prove that every positive integer is included in one of the sequences, let k be any natural number and $n = [(k+1)/\alpha]$. Since α is irrational, we have

$$n < \frac{k+1}{\alpha} < n+1, \quad n\alpha < k+1 < (n+1)\alpha.$$

If $n\alpha > k$, then $[n\alpha] = k$. If $n\alpha < k$, then

$$\frac{n}{k} < \frac{1}{\alpha} < \frac{n+1}{k+1}, \quad 1 - \frac{n}{k} > 1 - \frac{1}{\alpha} > 1 - \frac{n+1}{k+1}.$$

Since $1 - 1/\alpha = 1/\beta$, we get

$$\frac{m}{k} > \frac{1}{\beta} > \frac{m}{k+1}, \quad \text{where } m = k - n.$$

This shows that $[m\beta] = k$. Thus, $k = [n\alpha]$ or $m[\beta]$, that is, k is an element of one of the sequences.

22. Let α and β be positive real numbers such that the two infinite sequences

$$[\alpha], [2\alpha], [3\alpha], \ldots \quad \text{and} \quad [\beta], [2\beta], [3\beta], \ldots$$

together contain each natural number exactly once. Prove that α and β are irrational numbers such that $\frac{1}{\alpha} + \frac{1}{\beta} = 1$.

Solution If $0 < \alpha < 1$, then $[\alpha] = 0$. This disagrees with the fact that all the terms in the first sequence are positive integers. If $\alpha = 1$, then the first sequence contains all positive integers, and some of them will be repeated in the second. Hence, we must have $\alpha > 1$. Similarly, $\beta > 1$.

Let S be the set $\{1, 2, \ldots, N\}$, where N is a large positive integer. Every element of S belongs to one of the two sequences, and none is repeated. Since $\alpha > 1$ and $\beta > 1$, we see that the terms in both sequences are strictly increasing. Let

$$[n\alpha] \le N < [(n+1)\alpha], \quad [m\beta] \le N < [(m+1)\beta].$$

Then, S contains exactly n terms $[r\alpha], 1 \le r \le n$, of the first sequence. Similarly, it contains m terms of the second sequence. Hence, $N = m + n$. Moreover,

$$n\alpha < [n\alpha] + 1 \le N + 1, \quad (n+1)\alpha \ge [(n+1)\alpha] \ge N + 1,$$

and so, $n\alpha < N+1 \le (n+1)\alpha$. Similarly, $m\beta < N+1 \le (m+1)\beta$. It follows that

$$n < \frac{N+1}{\alpha} \le n+1, \quad m < \frac{N+1}{\beta} \le m+1.$$

Add these two. Since $N = m + n$, we get

$$N < (N+1)\delta \le N+2, \quad \text{where } \delta = 1/\alpha + 1/\beta,$$

and hence, $1 - 1/(N+1) < \delta \le 1 + 1/(N+1)$. This is true however large N is and implies that $\delta = 1$.

Thus, $1/\alpha + 1/\beta = 1$. If α or β is rational, the other is also rational. Let $\alpha = a/b$, where a, b are positive integers. Then $a > b$, and

$$\beta = \frac{\alpha}{\alpha - 1} = \frac{a}{a-b}, \quad b\alpha = a = (a-b)\beta.$$

Hence, both the sequences contain the number a. This shows that if no natural number belongs to both sequences, then α is irrational. Similarly, β is also irrational.

23. Let p be prime, k a positive integer, and $1 \le n \le p^k - 1$. Prove that

(a) $\binom{p^k}{n} \equiv 0 \pmod{p}$.

(b) If, further, n is not divisible by p^{k-1}, then $\binom{p^k}{n} \equiv 0 \pmod{p^2}$.

Solution We have $\binom{p^k}{n} = \dfrac{(p^k)\,!}{n\,!\,(p^k-n)\,!}$. This is known to be an integer. To see whether it is divisible by p or p^2, consider the highest powers of p that divide the numerator and the denominator.

Let p^e, p^f and p^g be the highest powers of p that divide $(p^k)\,!$, $n\,!$ and $(p^k-n)\,!$, respectvely. Then,

$$e = \sum_i \left[\frac{p^k}{p^i}\right], \quad f = \sum_i \left[\frac{n}{p^i}\right], \quad g = \sum_i \left[\frac{p^k - n}{p^i}\right].$$

If N is an integer, then $N - [x] \ge N - x > N - [x] - 1$, and so $[N-x] = N - [x] - 1$ unless x is an integer, and $[N-x] = N - [x]$ if x is an integer. This shows that for $1 \le i \le k$,

$$\left[\frac{p^k - n}{p^i}\right] = \begin{cases} p^{k-i} - \left[\frac{n}{p^i}\right] - 1 & \text{if } n \text{ is not divisible by } p^i, \\ p^{k-i} - \left[\frac{n}{p^i}\right] & \text{otherwise.} \end{cases}$$

Since n is not divisible by p^i when $i = k$, we see that

$$\left[\frac{n}{p^i}\right] + \left[\frac{p^k - n}{p^i}\right] = p^{k-i} - 1 \qquad (*)$$

for at least one value of i. Summing over i, we see that $f + g \leq e - 1$. Thus, $e \geq f + g + 1$, and so $\binom{p^k}{n}$ is divisible by p. This proves (a). If, further, n is not divisible by p^{k-1}, then n is not divisible by p^i when $i = k$ or $k - 1$. Hence, (*) holds for at least two values of i. It follows that $e \geq f + g + 2$, and so $\binom{p^k}{n}$ is divisible by p^2.

24. (a) Determine all positive integers n such that $[\sqrt{n}]$ divides n.
    (b) Determine all positive integers n, not a perfect square, such that $[\sqrt{n}]^3$ divides n^2.

Solution (a) Let $m = [\sqrt{n}]$ so that $m \leq \sqrt{n} < m + 1$. If $m|n$, let $n = am$. Then,

$$m^2 \leq n < (m+1)^2, \quad m \leq a < m + 2 + 1/m,$$

and hence, $a = m, m + 1$ or $m + 2$. The desired values of n are of the form m^2, $m(m+1)$, $m(m+2)$ for positive integers m.

(b) Let $m = [\sqrt{n}]$ so that $m < \sqrt{n} < m + 1$. If $m^3|n^2$, let $n^2 = am^3$. Then $(n/m)^2$ is an integer and so, n/m is an integer. Since n is not a perfect square, we see by (a) that (i) $n = m(m+1)$ or (ii) $n = m(m+2)$. In case (i),

$$am^3 = n^2 = m^2(m+1)^2, \quad am = (m+1)^2 = m^2 + 2m + 1.$$

This implies that $m = 1$, $n = m(m+1) = 2$. In case (ii),

$$am^3 = n^2 = m^2(m+2)^2, \quad am = (m+2)^2 = m^2 + 4m + 4.$$

This implies that $m|4$. Hence, $m = 1$, 2 or 4, and $n = m(m+2) = 1 \times 3, 2 \times 4$ or 4×6. Thus, the desired values of n are 2, 3, 8 and 24.

25. Prove that $[3x]+[3y] \geq [x]+[y]+2[x+y]$ for all real numbers x, y.

Solution Let $a=[x], b=[y], u=x-a$ and $v=y-b$. Then, $0 \leq u, v < 1$. Since $[N+t]=N+[t]$ for integers N, we see that

$$[3x]+[3y]=3a+[3u]+3b+[3v], \quad [x+y]=a+b+[u+v].$$

So the inequality to be proved reduces to

$$[3u]+[3v] \geq 2[u+v]. \tag{*}$$

Using the fact that $[\alpha+\beta]=[\alpha]+[\beta]$ or $[\alpha]+[\beta]+1$, for all real α, β, we get

$$[3u] \geq [u+v]+[2u-v], \quad [3v] \geq [u+v]+[2v-u],$$

$$[2u-v]+[2v-u] \geq [(2u-v)+(2v-u)]-1=[u+v]-1.$$

Combining these, we get

$$[3u]+[3v] \geq 2[u+v]+[u+v]-1.$$

Hence, (*) holds if $[u+v] \geq 1$. Note that $[u+v]=0$ or 1. If it is 0, then (*) is true, trivially.

26. Prove that $[4x]+[4y] \geq [x]+[y]+[2x+y]+[2y+x]$ for real numbers x, y.

Solution As above, let $a=[x], b=[y], u=x-a$ and $v=y-b$. Then, $0 \leq u, v < 1$. The inequality to be proved reduces to

$$[4u]+[4v] \geq [2u+v]+[2v+u]. \tag{*}$$

Because of the symmetry in u and v, we may suppose that $u \leq v$. First observe that if $[2u+v]=0$, then $[4u] \geq [2u+v]$. Since $[4v] \geq [2v+u]$, we see that (*) holds if $2u+v<1$. Since $[\alpha+\beta] \geq [\alpha]+[\beta]$ for all α, β, we have

$$[4u] \geq [2u+v]+[2u-v], \quad [4v] \geq [2v+u]+[2v-u].$$

Hence, (*) holds if $[2u-v]+[2v-u] \geq 0$. But $[2u-v] \geq -1$. So we need consider only the case in which

$$2u+v \geq 1, \quad 2u-v<0, \quad 2v-u<1.$$

Since $2u+v<2v+u=(2v-u)+(2u-v)+v<1+0+1$, we see that $[2u+v] \leq [2v+u] \leq 1$. So the right-hand side of (*) is at most 2.

If $u \geq 1/4$, then $[4v] \geq [4u] \geq 1$. On the other hand, if $u<1/4$, then $v \geq 1-2u>1/2$, and hence, $[4v] \geq 2$. Thus, (*) holds in any case.

27. Prove that $[5x] + [5y] \geq [x] + [y] + [3x + y] + [3y + x]$ for all real numbers x, y.

Solution As above, let $a = [x], b = [y], u = x - a$ and $v = y - b$. The inequality to be proved reduces to

$$[5u] + [5v] \geq [3u + v] + [3v + u]. \qquad (*)$$

Because of the symmetry in u and v, we may suppose that $u \leq v$. First observe that if $[3u+v] = 0$, then $[5u] \geq [3u+v]$. Since $[5v] \geq [3v+u]$, we see that (*) holds if $3u + v < 1$. Since $[\alpha + \beta] \geq [\alpha] + [\beta]$ for all α, β, we have

$$[5u] \geq [3u + v] + [2u - v], \quad [5v] \geq [3v + u] + [2v - u].$$

Hence, (*) holds if $[2u - v] + [2v - u] \geq 0$. But $[2u - v] \geq -1$. So we need consider only the case in which

$$3u + v \geq 1, \quad 2u - v < 0, \quad 2v - u < 1.$$

Since

$$3u + v = 2(2u - v) + (2v - u) + v < 0 + 1 + 1,$$
$$3v + u = (2u - v) + (2v - u) + 2v < 0 + 1 + 2,$$

we see that $[3u + v] \leq 1$, $[3v + u] \leq 2$. So the right-hand side of (*) is at most 3.

If $v \geq 3/5$, then $[5v] \geq 3$, and so (*) holds. So let $v < 3/5$. If $u \geq 1/5$, then $v > 2u \geq 2/5$, and hence $[5u] + [5v] \geq 1 + 2$. If $u < 1/5$, then

$$v \geq 1 - 3u > 1 - \frac{3}{5} = \frac{2}{5}, \quad 3u + v < 3v + u < \frac{9}{5} + \frac{1}{5} = 2.$$

Hence, $[5v] \geq 2$, while the right-hand side of (*) is at most $1 + 1$. Thus, (*) is true in any case.

28. Determine all positive integers k with the property that

$$[kx] + [ky] \geq [x] + [y] + [(k - 2)x + y] + [(k - 2)y + x]$$

for all real numbers x and y.

Solution We have proved that the inequality is true when $k = 3, 4$ and 5. When $k = 1$ and $k = 2$, this reduces, respectively, to

$$[x] + [y] \geq [x] + [y] + [-x + y] + [-y + x],$$
$$[2x] + [2y] \geq [x] + [y] + [y] + [x].$$

For all real numbers α, we have $[-\alpha] = -[\alpha]$ or $-[\alpha] - 1$, and $[2\alpha] = 2[\alpha]$ or $2[\alpha] + 1$, and hence, $[-\alpha] + [\alpha]$ is 0 or -1, and $[2\alpha] - 2[\alpha]$ is 0 or 1. Thus, the last two inequalities are true.

It can be proved that it is false when $k \geq 6$. For this, choose

$$x = \frac{1}{k} - \frac{1}{2k^2}, \quad y = \frac{3}{k} - \frac{1}{2k^2}.$$

Then $kx < 1$ and $ky < 3$. On the other hand,

$$(k-2)x + y = \frac{(k-2)+3}{k} - \frac{(k-2)+1}{2k^2} > 1,$$
$$(k-2)y + x = \frac{3(k-2)+1}{k} - \frac{(k-2)+1}{2k^2} > 2.$$

It follows that the left-hand side is not more than $0 + 2$, and that the right-hand side is not less than $1 + 2$. So the inequality is not true for these values of x, y.

29. Let m and n be positive integers. Prove that

$$\frac{(3m)!\,(3n)!}{m!\,n!\,(m+n)!\,(m+n)!}, \quad \frac{(4m)!\,(4n)!}{m!\,n!\,(2m+n)!\,(2n+m)!},$$
$$\frac{(5m)!\,(5n)!}{m!\,n!\,(3m+n)!\,(3n+m)!}$$

are integers, and that

$$\frac{(6m)!\,(6n)!}{m!\,n!\,(4m+n)!\,(4n+m)!}$$

is not necessarily an integer.

Solution It is known that the highest power of a prime p that divides $n!$ is p^e, where

$$e = \left[\frac{n}{p}\right] + \left[\frac{n}{p^2}\right] + \left[\frac{n}{p^3}\right] + \cdots.$$

Let $x = m/p^r$, $y = n/p^r$, where r is a positive integer. If $k = 3, 4$ or 5, then we have proved in the previous problem that

$$[kx] + [ky] \geq [x] + [y] + [(k-2)x + y] + [(k-2)y + x].$$

This readily shows that the index of the power of any prime p in the expression

$$\frac{(km)!\,(kn)!}{m!\,n!\,((k-2)m+n)!\,((k-2)n+m)!}, \qquad (*)$$

is non-negative, when $k = 3, 4$ or 5. Hence, this rational number is actually an integer.

For the second part, let $k = 6$, $m = 2$ and $n = 6$. Then, $(k-2)m+n = 14$, and $(k-2)n + m = 26$. So the expression in (*) becomes

$$\frac{12!36!}{2!6!14!26!} = \frac{36 \times 35 \times \cdots \times 27}{2!6!\,14 \times 13}.$$

This is not an integer since its numerator is not divisible by 13.

More generally, let $k \geq 6$. If m is such that $p = km + 1$ is prime, $m \geq 2$ and $n = 3m$, then the expression in (*) reduces to

$$\frac{(3km)(3km-1)\cdots(3km-5m+1)}{m!n!((km+m)(km+m-1)\cdots(km+1)}.$$

Since $3km < 3km + 3 = 3p$ and $3km - 5m + 1 > 2km + 2 = 2p$, we see that the numerator of this not divisible by p. Hence, this is not an integer.

30. Determine all integers $n > 1$ such that $(n-2)!$ is not divisible by n^2.

Solution If n is prime, then $(n-2)!$ is not divisible by n. Note that $0! = 1$. Thus, all primes have the desired property. So consider composite numbers n.

Let p be a prime divisor of n. The exponent of the highest power of p that divides $(n-2)!$ is

$$e_p = \left[\frac{n-2}{p}\right] + \left[\frac{n-2}{p^2}\right] + \left[\frac{n-2}{p^3}\right] + \cdots.$$

Since $n - 2 = p^i\left(\dfrac{n}{p^i} - 1\right) + (p^i - 2)$, we see that

$$\left[\frac{n-2}{p^i}\right] = \frac{n}{p^i} - 1, \quad \text{if } p^i | n,\ i \geq 1.$$

First consider the case in which n has only one prime divisor p. Then $n = p^r$ for some $r \geq 2$, and $[(n-2)/p^i] = n/p^i - 1 = p^{r-i} - 1$ for $1 \leq i \leq r$. Thus,

$$e_p = p^{r-1} + p^{r-2} + \cdots + 1 - r.$$

Let $r = 2$ so that $n = p^2$. Then, $e_p = p - 1$. So if $p \geq 5$, then $e_p \geq 4$ which implies that $(n-2)!$ is divisible by $p^4 = n^2$. If $p = 2$, then $n = 4$, and if $p = 3$ then $n = 9$. Since 2! is not divisible by 4^2 and 7! is not divisible by 9^2, we see that 4 and 9 have the desired property.

Let $r = 3$ so that $n = p^3$. Then, $e_p = p^2 + p - 2 \geq 10$ if $p \geq 3$. This implies that $(n-2)!$ is divisible by $p^6 = n^2$ if $n = p^3, p \geq 3$. If $p = 2$, then $n = 8$. Since 6! is not divisible by 8^2, we see that 8 has the desired property.

Let $r \geq 4$. If $m \geq 2$, then

$$2^m = (1+1)^m = 1 + \binom{m}{1} + \cdots + \binom{m}{m} \geq m + 2.$$

This shows that

$$p^{r-2} \geq 2^{r-2} \geq r, \quad p^{r-1} \geq 2^{r-1} \geq 2r.$$

Consequently, $e_p \geq p^{r-1} + p^{r-2} - r \geq 3r - r = 2r$. This implies that $(n-2)!$ is divisible by $p^{2r} = n^2$, that is, n does not have the desired property.

Next consider the case in which n has exactly two prime divisors p and q. Then $n = p^r q^s$, and so $n = ap^r$, where $a = q^s$, $r \geq 1$, and $n = bq^s$, where $b = p^r$, $s \geq 1$.

Since

$$\left[\frac{n-2}{p^i}\right] = \frac{n}{p^i} - 1 = ap^{r-i} - 1 \geq a - 1 \quad \text{for } 1 \leq i \leq r, \qquad (*)$$

we first observe that $e_p \geq (a-1)r$. Similarly, $e_q \geq (b-1)s$. It follows that if $a - 1 \geq 2$ and $b - 1 \geq 2$, then $(n-2)!$ is divisible by p^{2r} and q^{2s}, and hence, by n^2. But $a - 1 < 2$ only when $q^s = a = 2$, that is, $q = 2$ and $s = 1$; similarly, $b - 1 < 2$ only when $p = 2$ and $r = 1$.

Suppose that $q = 2$ and $s = 1$. Then $n = 2p^r$, where $p \geq 3$. First suppose that $r = 1$ so that $n = 2p$. Then $e_p = n/p - 1 = 1$. So $(n-2)!$ is not divisible by p^2.

Next let $r \geq 2$. It follows from (*) with $a = 2$ that

$$\begin{aligned} e_p &= (2p^{r-1} - 1) + (2p^{r-2} - 1) + \cdots \\ &= 2\left(p^{r-1} + p^{r-2} + \cdots\right) - r. \end{aligned}$$

If $m \geq 1$, then

$$3^m = (1+2)^m = 1 + \binom{m}{1} 2 + \cdots + \geq 2m + 1.$$

Hence,

$$p^{r-1} \geq 3^{r-1} \geq 2(r-1) + 1 = 2r - 1.$$

Thus, we see that $e_p \geq 2(2r-1) - r \geq 2r$. This implies that $(n-2)!$ is divisible by p^{2r}. Now, $e_2 = n/2 - 1 = p^r > 2$, which implies that $(n-2)!$ is divisible by 2^2. It follows that $(n-2)!$ is divisible by n^2. Thus, if $n = 2p^r$, $p \geq 3$, then n has the desired property if and only if $r = 1$. The same argument applies to the case $p = 2$, $r = 1$.

Finally, consider the case in which n has at least three distinct prime divisors, and suppose that p is one of them. Then $n = ap^r$, where $r \geq 1$. From (*) we see that $e_p \geq (a-1)r$. But $a \geq 3$ since a has at least two prime divisors. Hence, $(n-2)!$ is divisible by p^{2r}. This argument can be applied to any prime divisor of n. It follows that $(n-2)!$ is divisible by n^2.

Combining these facts, we conclude that $(n-2)!$ is not divisible by n^2 if (i) $n = p$ or $2p$, where p is prime, or (ii) $n = 8$ or 9. If n is any other integer greater than 1, then $(n-2)!$ is divisible by n^2.

Set III

1. Prove that $\frac{1}{2}\sqrt{n} \leq \phi(n) \leq n$ for all n.

   Solution Since $\phi(1) = 1$ and $\phi(n) \leq n - 1$ for $n > 1$, it is enough if we prove the first inequality. Let $n = 2^a m$, where m is odd. If $m > 1$, let $m = \prod_i p_i^{k_i}$, where p_i are odd primes and $k_i \geq 1$. Since $p - 1 > \sqrt{p}$ for $p \geq 3$, we get

$$\phi(m) = \prod_i p_i^{k_i - 1}(p_i - 1) \geq \prod_i p_i^{k_i - 1}\sqrt{p_i} = \prod_i p_i^{k_i - \frac{1}{2}},$$

But $k - \frac{1}{2} \geq k/2$ for $k \geq 1$. Hence,

$$\phi(m) \geq \prod_i p_i^{k_i/2} = \sqrt{m}.$$

This holds even when $m = 1$.

Since $\phi(2^a) = 2^{a-1}$ if $a \geq 1$, and $\phi(2^a) = 1$ if $a = 0$, we see that

$$2\phi(n) = 2\phi(2^a)\phi(m) \geq 2^a\sqrt{m} \geq \sqrt{2^a m} = \sqrt{n}.$$

2. Prove that for a fixed integer n, the equation $\phi(x) = n$ has only a finite number of solutions.

Solution If N is a positive integer such that $\phi(N) = n$, then $n \geq \frac{1}{2}\sqrt{N}$ by the previous problem. There can be at the most a finite number of values of N satisfying this.

3. Prove that

(a) if n has r distinct prime divisors, then $\phi(n) \geq n/2^r$,

(b) if n has r distinct odd prime divisors, then $\phi(n)$ is divisible by 2^r.

Solution Let $n = \prod_{i=1}^{r} p_i^{n_i}$, where p_i are distinct primes and $n_i \geq 1$. Then,

$$\phi(n) = \prod_{i=1}^{r} p_i^{n_i - 1}(p_i - 1).$$

Hence,

$$2^r\phi(n) = \prod_{i=1}^{r} p_i^{n_i-1} \cdot 2(p_i - 1) \geq \prod_{i=1}^{r} p_i^{n_i-1} \cdot p_i = n,$$

since $2(m-1) \geq m$ for all $m \geq 2$. This proves (a).

If n has r distinct odd prime divisors, then the product for $\phi(n)$ contains r even factors $p_i - 1$. This readily gives (b).

4. Prove that there is no solution for the equation $\phi(x) = 14$, and that 14 is the least even integer with this property. Apart from 14, what is the next smallest even integer n such that $\phi(x) = n$ has no solution?

Solution If $2p + 1$ is prime, then $\phi(2p+1) = 2p$. Thus, the equation $\phi(x) = 2p$ has a solution if $2p + 1$ is prime.

Suppose, on the other hand, that the equation $\phi(x) = 2p$ has a solution, where p is an odd prime. Let $\phi(N) = 2p$. If N has r distinct odd prime divisors, then $\phi(N)$ is divisible by 2^r. This shows that $r \leq 1$, and so N has at most one odd prime divisor q. Let $N = 2^s q^a$, where $s \geq 0$, $a \geq 0$. Then,

$$2p = \phi(N) = \phi(2^s)\phi(q^a).$$

This implies that p divides $\phi(2^s)$ or $\phi(q^a)$. Since $\phi(2^s) = 1$ or 2^{s-1} and p is an odd prime, we see that p does not divide $\phi(2^s)$. Hence, $p|\phi(q^a)$, and so we cannot have $a = 0$. It follows that

$$2p = \phi(q^a) = q^{a-1}(q-1), \quad p = q^{a-1}\frac{q-1}{2}.$$

Consequently, $(q-1)/2 = 1$ or p, that is, $q = 3$ or $q = 2p+1$. If $q = 3$, we get $p = q^{a-1} = 3^{a-1}$, which implies that $p = 3$ and $2p + 1 = 7$. Thus, $2p + 1 = 7$ or q, which is prime in any case.

We proved that if p is prime, then $\phi(x) = 2p$, has a solution only if $2p + 1$ is prime. Hence, the equations $\phi(x) = 14$ and $\phi(x) = 26$ have no solution since 15 and 27 are not primes. If $n + 1$ is prime, then $\phi(x) = n$ has a solution, namely $x = n + 1$. Note that $n + 1$ is prime when $n = 2, 4, 6, 10, 12, 16, 18, 22$. Further, $\phi(16) = 8$, $\phi(25) = 20$ and $\phi(35) = 24$. Hence, 14 and 26 are the two smallest even integers such that the equation $\phi(x) = n$ has no solution.

5. Find all positive integers n for which $\phi(n) \equiv 2 \pmod 4$.

Solution First observe that $\phi(n) \equiv 2 \pmod 4$ if and only if $\phi(n)/2$ is odd. If n has an odd prime divisor p, let $n = p^a m$ where m is not divisible by p. Then $\phi(n)/2 = p^{a-1}\left((p-1)/2\right)\phi(m)$. This is odd only if $(p-1)/2$ is odd and $\phi(m) = 1$, that is, $(p-1)/2 = 2k+1$ and $m = 1$ or 2. So $p = 4k + 3$. If n has no odd prime divisor, let $n = 2^r$. Then $\phi(n)/2 = 2^{r-2}$ is odd if and only if $r = 2$, that is, $n = 4$. Thus, the desired values of n are 4, p^a, $2p^a$, where $a \geq 1$, p is prime and $p \equiv 3 \pmod 4$.

6. Prove that, for $n \geq 2$, the sum of all positive integers less than n and relatively prime to n is $\frac{1}{2}n\phi(n)$.

Solution Let $a_1, a_2, \ldots, a_k$, where $k = \phi(n)$, be the positive integers less than n and relatively prime to n. If a is prime to n, so is $n - a$. Hence, $n - a_1, n - a_2, \ldots, n - a_k$ are the same numbers written in another order. Consequently, if $S = \sum_i a_i$, then $S = \sum_i (n - a_i) = kn - S$, $2S = kn$, $S = \frac{1}{2}kn = \frac{1}{2}n\phi(n)$.

7. Define $f(n)$ as the sum of the positive integers less than n and prime to n. Prove that $f(m) = f(n)$ implies that $m = n$.

Solution By the previous problem, $f(n) = \frac{1}{2}n\phi(n)$. Hence, if $f(m) = f(n)$, then $m\phi(m) = n\phi(n)$, that is,

$$m^2 \prod_{p|m}(1 - 1/p) = n^2 \prod_{p|n}(1 - 1/p).$$

Cancel the common factors. Denote the primes, if any, dividing m but not n, by $p_1, \ldots, p_k$, where $p_1 < \cdots < p_k$. Let those dividing n but not m, be $q_1, \ldots, q_r$, where $q_1 < \cdots < q_r$. We get $m^2 \prod_{i=1}^{k}(1 - 1/p_i) = n^2 \prod_{j=1}^{r}(1 - 1/q_j)$. If $m = m_1 p_k$, this can be rearranged as

$$m_1^2 p_k \prod_{i=1}^{k}(p_i - 1) \prod_{j=1}^{r} q_j = n^2 \prod_{j=1}^{r}(q_j - 1) \prod_{i=1}^{k-1} p_i.$$

Since p_k does not divide n, and $p_k > p_i$ for $i < k$, this implies that $p_k | q_j - 1$ for some j. A similar argument shows that $q_r | p_i - 1$ for some i. Thus, $p_k \le q_j - 1 \le q_r - 1$, and $q_r \le p_i - 1 \le p_k - 1$. This contradiction shows that there are no such primes p_i, q_j. We are left with $m^2 = n^2$.

8. Determine all positive integers n such that

(a) $\phi(n) = n/2$,

(b) $\phi(n) = n/3$.

Solution (a) If $n = 2^r, r \ge 1$, then $\phi(n) = 2^{r-1} = n/2$. Suppose, on the other hand, that $\phi(n) = n/2$. Since $\phi(n) = n \prod_{p|n}(1 - 1/p)$, we see that if $\phi(n) = n/2$, then $\prod_{p|n}(1 - 1/p) = 1/2$, that is,

$$2 \prod_{p|n}(p - 1) = \prod_{p|n} p.$$

So one of the primes dividing n must be 2. If n has also an odd prime divisor, then the product on the left-hand side is divisible by 4. This shows that n cannot have an odd prime divisor. Thus, $\phi(n) = n/2$ if and only if n is of the form $n = 2^r, r \ge 1$.

(b) Similarly, $\phi(n) = n/3$ only if

$$3 \prod_{p|n}(p - 1) = \prod_{p|n} p.$$

Then one of the primes dividing n must be 3, and the left-hand side contains the factor $3-1=2$. Hence, one of the primes dividing n must be 2. So the right-hand side is $2m$ for some odd integer m. If n has one more odd prime divisor, then the left-hand side contains two even factors. This is not possible. Thus, $\phi(n)=n/3$ only if n is of the form $n=2^r3^s$, $r\geq 1$, $s\geq 1$. On the other hand, if n is of this form, then

$$\phi(n)=2^{r-1}(2-1)\,3^{s-1}(3-1)=2^r3^{s-1}=\frac{n}{3}.$$

9. Prove that if k is an integer greater than 3, then there are no integers n for which $\phi(n)=n/k$.

Solution Let $\phi(n)=n/k$, where $k>3$. Then $\prod_{p|n}(1-1/p)=1/k$, that is

$$k\prod_{p|n}(p-1)=\prod_{p|n}p. \tag{*}$$

If n has no odd prime divisor, then $n=2^r$ and $\phi(n)=n/2$. Hence, n has an odd prime divisor p, and the product on the left-hand side of (*) contains at least one even factor $p-1$. Consequently, one of the primes dividing n must be 2. Assume that n has at least two odd prime divisors. Then the left-hand side of (*) contains at least two even factors, while the right-hand side is of the form $2m$ for some odd m. So (*) does not hold if n has more than one odd prime divisor. It follows that n is of the form $n=2^rp^s$, where p is an odd prime, $r\geq 1$, $s\geq 1$. In that case, (*) becomes

$$k(2-1)(p-1)=2p,\quad p=k\frac{p-1}{2}.$$

Since $(p-1)/2<p$, this implies that $(p-1)/2=1$, which forces $k=p=3$. Then, $n=2^r3^s$, and $\phi(n)=n/3$. This shows that if $\phi(n)=n/k$ for some n, then $k=2$ or 3.

10. Find all integers n for which 2^n-1 is divisible by 7. Also show that there is no positive integer n for which 2^n+1 is divisible by 7.

Solution Any positive integer n is of the form $3k$, $3k+1$ or $3k+2$. Correspondingly, $2^n=8^k, 2\cdot 8^k$ or $4\cdot 8^k$. Since $8^k\equiv 1 \pmod 7$, we respectively have $2^n-1\equiv 0$, 1 or 3, modulo 7, and $2^n+1\equiv 2$, 3 or 5, modulo 7. This proves the second assertion. The positive integers n for which 2^n-1 is divisible by 7 are those of the form $3k$.

11. Let n be a positive integer which is not a square. Prove that for every integer a relatively prime to n, there exist integers x and y satisfying

$$ax \equiv y \pmod{n} \quad \text{with } 0 < |x| < \sqrt{n} \text{ and } 0 < |y| < \sqrt{n}.$$

Solution Let $m = [\sqrt{n}]$. Then $m^2 < n < (m+1)^2$. Consider the set

$$S = \{ra + s : 0 \le r \le m, 0 \le s \le m\}.$$

There are $(m+1)^2$ members in S, and $(m+1)^2 > n$. Hence, two of them are congruent modulo n. Suppose that $r_1 a + s_1 \equiv r_2 a + s_2 \pmod{n}$, where $r_1 \ne r_2$ or $s_1 \ne s_2$, and $0 \le r_i, s_i \le m$. Then,

$$(r_1 - r_2)a \equiv s_2 - s_1 \pmod{n}, \quad |r_1 - r_2| \le m, \quad |s_1 - s_2| \le m.$$

So, if $x = r_1 - r_2$ and $y = s_2 - s_1$, we get $ax \equiv y \pmod{n}$, where $|x| \le m < \sqrt{n}$ and $|y| \le m < \sqrt{n}$. If $x = 0$, then $y \equiv 0 \pmod{n}$, and hence $y = 0$. On the other hand, since $(a, n) = 1$,

$$y = 0 \Longrightarrow ax \equiv 0 \pmod{n} \Longrightarrow x \equiv 0 \pmod{n} \Longrightarrow x = 0.$$

But if $x = 0$ and $y = 0$, then $r_1 = r_2$ and $s_1 = s_2$. It follows that $x \ne 0$ and $y \ne 0$. This proves the result.

12. Let p be an odd prime. Prove the following:

(a) If $q = (p-1)/2$, then $(q\,!)^2 + (-1)^q \equiv 0 \pmod{p}$.

(b) $1^2 3^2 5^2 \cdots (p-2)^2 \equiv (-1)^{(p+1)/2} \pmod{p}$.

(c) $2^2 4^2 6^2 \cdots (p-1)^2 \equiv (-1)^{(p+1)/2} \pmod{p}$.

Solution Let $a = q\,!$. Since $p - 1 = 2q$, we have

$$(p-1)\,! = (2q)\,! = a(q+1)(q+2)\cdots(2q).$$

But $q + r = p - 1 - q + r \equiv -(q - r + 1) \pmod{p}$. Using this with $r = 1, 2, \ldots, q$, we get

$$(p-1)\,! \equiv a(-1)^q (q\,!) \pmod{p} \equiv (-1)^q a^2 \pmod{p}.$$

Thus, $(-1)^q a^2 + 1 \equiv (p-1)\,! + 1 \equiv 0 \pmod{p}$, by Wilson's theorem. It follows that $a^2 + (-1)^q \equiv 0 \pmod{p}$. This proves (a).

Multiplying the congruences $2r \equiv -(p-2r) \pmod{p}$ for $1 \le r \le 2q$, we get

$$2 \cdot 4 \cdots (2q) \equiv (-1)^q (p-2)(p-4) \cdots 1 \pmod{p}.$$

Multiply both sides by $1 \cdot 3 \cdots (p-2)$ and $2 \cdot 4 \cdots (2q)$. We get

$$(p-1)! \equiv (-1)^q 1^2 \cdot 3^2 \cdots (p-2)^2 \pmod{p},$$
$$2^2 \cdot 4^2 \cdots (p-1)^2 \equiv (-1)^q (p-1)! \pmod{p},$$

respectively. Since $(p-1)! \equiv -1 \pmod{p}$, we get

$$1^2 \cdot 3^2 \cdots (p-2)^2 \equiv (-1)^q(-1) \pmod{p},$$
$$2^2 \cdot 4^2 \cdots (p-1)^2 \equiv (-1)^q(-1) \pmod{p},$$

proving (b) and (c) as $q+1 = (p+1)/2$.

13. Let p be a prime and $p \equiv 1 \pmod{4}$. Prove that there exist positive integers x and y such that $p = x^2 + y^2$. Show also that such an expression of p is unique except for the order of summands.

Solution Let $p \equiv 1 \pmod{4}$, $q = (p-1)/2$ and $a = q!$. Then q is even, and $(a, p) = 1$. So by the previous problem, we have

$$a^2 + 1 = (q!)^2 + 1 \equiv 0 \pmod{p}.$$

Further, there exist integers x and y such that

$$ax \equiv y \pmod{p}, \quad \text{where } 0 < |x| < \sqrt{p},\ 0 < |y| < \sqrt{p}.$$

Hence, $y^2 \equiv a^2x^2 \equiv -x^2 \pmod{p}$; that is, $x^2 + y^2 \equiv 0 \pmod{p}$. Let $x^2 + y^2 = mp$. Since $0 < x^2 + y^2 < p + p$, we have $0 < m < 2$, that is, $m = 1$. Thus, $x^2 + y^2 = p$. This proves the existence.

Uniqueness: Let $x^2 + y^2 = p = u^2 + v^2$. Then,

$$y^2u^2 \equiv -y^2v^2 \equiv x^2v^2 \pmod{p}.$$

Thus, $(yu - xv)(yu + xv)$ is divisible by p. Hence, one of the factors must be divisible by p. First suppose that $yu - xv$ is divisible by p. Since $x^2 + y^2 = p = u^2 + v^2$, we see that each of x, y, u, v is less than $\sqrt{p}$. Hence, $0 < yu < p$ and $0 < xv < p$. This implies that $yu = xv$. Since $(u, v) = 1$, we get $u|x$, and since $(x, y) = 1$, we get $x|u$. Thus, $x = u$ and so $y = v$.

On the other hand, if $yu + xv$ is divisible by p, then $yu \equiv -xv \pmod{p}$,

$$xyu \equiv -x^2v \equiv y^2v \pmod{p}, \quad xu \equiv yv \pmod{p}.$$

As before, this implies that $xu = yv$ which forces $x = v$, $y = u$.

14. Consider a prime $p \equiv 1 \pmod{60}$. Show that there exist positive integers a and b with $p = a^2 + b^2$, where 3 divides a or b and 5 divides a or b.

Solution It is known that any prime of the form $4k+1$ is expressible as the sum of two squares. Here $p = 60n + 1$ is of this form. So we can find a and b. Suppose that 3 does not divide a. Then $a^2 \equiv 1 \pmod{3}$, and hence $60n + 1 \equiv 1 + b^2 \pmod{3}$, implying that b is divisible by 3.

Next suppose that 5 does not divide a. Then $a^2 \equiv 1$ or 4 modulo 5, and hence $60n + 1 \equiv 1 + b^2 \pmod{5}$ or $4 + b^2 \pmod{5}$. Thus, $b^2 \equiv 0$ or 2 modulo 5. The latter is impossible. So b is divisible by 5.

15. Let the positive integer n be written as $n = N^2m$, where m is square-free. Prove that n can be represented as the sum of two squares if and only if m has no prime divisor of the form $4k + 3$.

Solution Suppose that m has no prime divisor of the form $4k+3$. Then $m = 1$ or it is a product of primes. Since $1 = 1^2 + 0^2, 2 = 1^2 + 1^2$, and a prime of the form $4k+1$ is the sum of two squares, we see that m is a product of factors, each of which is a sum of two squares. The identity

$$(a^2 + b^2)(c^2 + d^2) = (ac + bd)^2 + (ad - bc)^2$$

shows that if m_1 and m_2 are sums of two squares, then m_1m_2 is also a sum of two squares. Repeated application of this shows that m is a sum of two squares. If $m = x^2 + y^2$, then $n = (Nx)^2 + (Ny)^2$, a sum of two squares.

Conversely, suppose that n is expressible as a sum of two squares. Let $n = a^2 + b^2$, $d = \gcd(a, b)$, and $a = dr$, $b = ds$ so that $\gcd(r, s) = 1$. Let p be an odd prime divisor of m. The highest power of p that divides p is p itself as m is square-free. Consider the highest powers of p that divide n, d, $r^2 + s^2$ and n. Let

$$n = p^e n_1,\ d = p^{e_1} d_1,\ r^2 + s^2 = p^{e_2} r_1,\ N = p^{e_3} N_1,$$

where none of n_1, d_1, r_1, N_1 is divisible by p, and $e, e_i \geq 0$. Then, $e = 2e_3 + 1$ as $n = N^2 m$, and so e is odd. On the other hand, $e = 2e_1 + e_2$ as $n = d^2(r^2+s^2)$. Hence, $e_2 \neq 0$ since e is odd. Consequently, r^2+s^2 is divisible by p. Note that r is divisible by p if and only if s is divisible by p. Since $\gcd(r, s) = 1$, this implies that r and s are relatively prime to p. Since $r^2 \equiv -s^2 \pmod{p}$, we get

$$(r^2)^{(p-1)/2} \equiv (-1)^{(p-1)/2} (s^2)^{(p-1)/2} \pmod{p},$$

that is, $r^{p-1} \equiv (-1)^{(p-1)/2} s^{p-1} \pmod{p}$. But $r^{p-1} \equiv 1 \equiv s^{p-1} \pmod{p}$. Thus, $(-1)^{(p-1)/2} \equiv 1 \pmod{p}$. This holds only when $(p-1)/2$ is even, that is, p is of the form $4k+1$.

16. Let $n = a^2+b^2$, where a and b are positive integers with $\gcd(a, b) = 1$. Prove that any positive divisor of n is a sum of two squares. Deduce that there are infinitely many primes of the form $4k+1$.

Solution Suppose that p is an odd prime divisor of $n = a^2 + b^2$. If a is divisible by p, then b is also divisible by p, which is not possible if $(a, b) = 1$. Hence, neither a nor b is divisible by p. Since $a^2 \equiv -b^2 \pmod{p}$, we get

$$(a^2)^{(p-1)/2} \equiv (-1)^{(p-1)/2} (b^2)^{(p-1)/2} \pmod{p}.$$

Assume that p is of the form $4k+3$. Then $(p-1)/2 = 2k+1$ is odd, and hence, $a^{p-1} \equiv -b^{p-1} \pmod{p}$. But $a^{p-1} \equiv 1 \equiv b^{p-1} \pmod{p}$, that is, $1 \equiv -1 \pmod{p}$. This is possible only when $p = 2$. It follows that n is a product of primes, each of which is either 2 or is of the form $4k+1$. Let d be any positive divisor of n. Then either $d = 1$ or d is a product of primes, each of which is 2 or of the form $4k+1$. Since $1 = 1^2 + 0^2$, $2 = 1^2 + 1^2$, and a prime of the form $4k+1$ is the sum of two squares, we see that d is a product of factors, each of which is a sum of two squares. The relation

$$(x^2 + y^2)(u^2 + v^2) = (xu + yv)^2 + (xv - yu)^2$$

shows that if m_1 and m_2 are sums of two squares, then $m_1 m_2$ is also a sum of two squares. Repeated application of this shows that d is a sum of two squares.

For the second part, assume that there are only a finite number of primes of the form $4k+1$. Let a be the product of all primes of the form $4k+1$, $b = 2$, and $n = a^2 + b^2$. Then n must have a prime divisor p. Clearly, p

cannot be any of the prime factors of a. But p is of the form $4k+1$ by the first part. This gives a contradiction if a is the product of *all* primes of this form.

17. Given an odd integer $n \geq 3$, let k and t be the least positive integers such that both $kn+1$ and tn are perfect squares. Prove that n is prime if and only if both k and t are greater than $n/4$.

Solution Note that we can always find k and t such that $kn+1$ and tn are perfect squares; for example, we can take $k = n+2$ and $t = n$. Our interest is in the least possible values for k and t.

Let n be an odd prime, $tn = x^2$ and $kn+1 = y^2$ for some positive integers x, y. Then n divides x and $y^2 - 1 = (y+1)(y-1)$. So $n|y+1$ or $n|y-1$. It follows that $n \leq x$ and $n \leq y+1$. Thus,

$$kn+1 = y^2 \geq (n-1)^2 = n^2 - 2n + 1, \quad k \geq n-2 > \frac{n}{4},$$
$$tn = x^2 \geq n^2, \quad t \geq n.$$

Hence, both k and t are greater than $n/4$ if n is an odd prime.

For the converse, assume that n is an odd composite number. We can write $n = ab$, where $a = p^r$ for an odd prime p, and b is odd and not divisible by p. First suppose that $b = 1$. Then $n = p^r$, $r \geq 2$. If r is even, let $t = 1$, and if r is odd, let $t = p$. In either case, tn is a square. Further, if r is odd, then $r \geq 3$, and

$$tn = p^{r+1} = \frac{n^2}{p^{r-1}} \leq \frac{n^2}{p^2} < \frac{n^2}{4}.$$

Hence, $t < n/4$, whether r is even or odd.

Next suppose that $b > 1$. Since $\gcd(a, b) = 1$, the congruence $ax \equiv 2 \pmod{b}$ has a solution. Since $b > 1$ and is odd, we can choose x with $1 \leq x < b$. Let $ax = by + 2$. There are two possibilities: (i) $0 < x < b/2$ and (ii) $b/2 < x < b$. In case (i), we have

$$nxy = (ax)(by) = (ax)(ax-2) = (ax-1)^2 - 1.$$

Thus, $kn+1 = (ax-1)^2$, where $k = xy$. Since $ax - 1 \geq a - 1 > 1$, we see that $k > 0$. In case (i), we also have $ax - 2 < ax < n/2$, and hence, $nxy < (n/2)(n/2)$. Thus, $k = xy < n/4$.

Next consider case (ii). Since b is odd, we have $(b+1)/2 \le x < b$. Let $x_1 = b - x$. Then, $0 < x_1 \le (b-1)/2$,

$$ax_1 + 2 = ab - ax + 2 = ab - by,$$
$$(ax_1+1)^2 - 1 = ax_1(ax_1+2) = ax_1\, b(a-y) = kn,$$

where $k = x_1(a-y)$. As before, we see that $k > 0$. Moreover,

$$ax_1 + 1 \le \frac{a(b-1)}{2} + 1 < \frac{ab}{2},$$

since $1 < a/2$. This shows that $kn + 1 = (ax_1+1)^2 < (n/2)^2$, and so $k < n/4$.

We proved that if n is an odd composite number and $n > 1$, either tn is a perfect square with some $t < n/4$, or else, $kn+1$ is a perfect square with some $k < n/4$. This completes the proof.

18. Let s_n denote the sum of the first n primes. Prove that, for each n, there is an integer whose square lies between s_n and s_{n+1}.

Solution Let $m = [\sqrt{s_n}\,]$ and p_n be the nth prime. Thus, $m^2 \le s_n < (m+1)^2$, and $p_1 = 2, p_2 = 3, \ldots$. If $p_n = 2k-1$, then the sum of all odd positive integers not exceeding p_n is $1+3+\cdots+2k-1 = k^2$. The sum of the primes among them is $s_n - 2$. If $n \ge 5$, then $p_n \ge 11$ and 9 is a composite number among them. It follows that $k^2 \ge 1+9+(s_n-2)$, and hence, $s_n < k^2 = ((p_n+1)/2)^2$ for all $n \ge 5$. This implies that $m^2 < ((p_n+1)/2)^2$, $m < (p_n+1)/2$ and $2m-1 < p_n$. Thus, we see that $p_{n+1} \ge p_n + 2 > 2m+1$, and

$$s_{n+1} = s_n + p_{n+1} > m^2 + 2m + 1.$$

Hence, for $n \ge 5$, we have $s_n < (m+1)^2 < s_{n+1}$, where $m = [\sqrt{s_n}\,]$. Since $s_1 = 2$, $s_2 = 5$, $s_3 = 10$, $s_4 = 17$ and $s_5 = 28$, we can see that this holds for all $n \ge 1$.

19. Find all positive integers n for which $(n-1)! + 1$ is a power of n.

Solution Suppose that $n > 5$, and let $(n-1)! + 1 = n^r$ for some $r > 1$. Then $(n-1)! = n^r - 1$, and so

$$\begin{aligned}(n-2)! &= n^{r-1} + n^{r-2} + \cdots + n + 1\\ &\equiv 1 + 1 + \cdots + 1 \pmod{n-1},\end{aligned}$$

that is, $(n-2)! \equiv r \pmod{n-1}$. Note that $2 < (n-1)/2 < n-2$ since $n > 5$. Thus, 2 and $(n-1)/2$ are two distinct factors in $(n-2)!$. It follows that $(n-2)! \equiv 0 \pmod{n-1}$, that is, $r \equiv 0 \pmod{n-1}$. Now, $(n-1)! < n^{n-2}$. This implies that $n^r = (n-1)! + 1 \le n^{n-2}$, $r \le n-2$. It follows that $r = 0$, which gives a contradiction.

For $n = 1, 2, 3, 4$ and 5, we have $(n-1)! + 1 = 2, 2, 3, 7$ and 25, respectively. Thus, we see that $(n-1)! + 1$ is a power of n if and only if $n = 2, 3$ and 5.

20. Prove that if n is a positive integer such that the equation

$$x^3 - 3xy^2 + y^3 = n$$

has a solution (x, y) in integers, then it has at least three such solutions. Show also that if n is congruent modulo 9 to 2, 4, 5 or 7, then the equation has no solution in integers.

Solution Suppose that (x, y) is a solution. We can write

$$\begin{aligned} x^3 - 3xy^2 + y^3 &= (y-x)^3 - 3x^2y + 2x^3 \\ &= (y-x)^3 - 3(y-x)x^2 - x^3. \end{aligned}$$

Hence, if $x_1 = y - x$ and $y_1 = -x$, then (x_1, y_1) is also a solution. If $x_1 = x$ and $y_1 = y$, then $y - x = x$ and $-x = y$, which forces $x = 0$, $y = 0$, $n = 0$. Since n is given to be non-zero, we see that (x_1, y_1) is different from (x, y).

The same argument shows that if $x_2 = y_1 - x_1 = -y$ and $y_2 = -x_1 = x - y$, then (x_2, y_2) is a solution different from (x_1, y_1). Since $y_2 - x_2 = x$ and $-x_2 = y$, we also find that (x, y) is different from (x_2, y_2). Thus, we get at least three different solutions (x, y), (x_1, y_1), (x_2, y_2).

Consider any positive integer $n \equiv 2, 4, 5 \text{ or } 7 \pmod{9}$. Suppose that (x, y) is a solution. If x is divisible by 3, then

$$n = x^3 - 3xy^2 + y^3 \equiv y^3 \pmod{9}.$$

But $y^3 \equiv 0, 1 \text{ or } 8 \pmod{9}$ for any integer y. Hence, in the present case, we must have $x \equiv 1 \text{ or } 2 \pmod{3}$. Similarly, we can see that $y \equiv 1 \text{ or } 2 \pmod{3}$. Thus, (i) $x \equiv y \pmod{3}$ or (ii) $x + y \equiv 0 \pmod{3}$. In case (i), define (x_1, y_1) as before. We get a solution (x_1, y_1) with $x_1 \equiv 0 \pmod{3}$. This is not possible. For case (ii), we have

$$n = x^3 - 3xy^2 + y^3 \equiv x^3 + y^3 \equiv x + y \equiv 0 \pmod{3}.$$

This too is impossible. Hence, the equation has no solution in integers if $n \equiv 2, 4, 5$ or $7 \pmod 9$.

21. Let p be a prime, $p \geq 5$ and write

$$1 + \frac{1}{2} + \frac{1}{3} + \cdots + \frac{1}{p} = \frac{r}{ps}.$$

Prove that $p^3 | (r - s)$.

Solution Problem 23 of Set VI states that if

$$1 + \frac{1}{2} + \frac{1}{3} + \cdots + \frac{1}{p-1} = \frac{m}{n}.$$

where m is relatively prime to n, then $p^2 | m$. So $m = Np^2$ for some integer N, n is not divisible by p, and

$$\frac{m}{n} = \frac{r}{ps} - \frac{1}{p} = \frac{r-s}{ps}, \quad mps = n(r-s)$$

. Hence, $n(r - s) = Np^3 s$ is divisible by p^3. This proves the result since p does not divide n.

22. If p is a prime, prove that $\binom{n}{p} \equiv \left[\frac{n}{p}\right] \pmod p$. Prove also that if p^a divides at least one of $\binom{n}{p}$ and $[n/p]$, then it divides the other also.

Solution Let $m = [n/p]$. Then $n = mp + r$, where $0 \leq r < p$. The p numbers $n, n - 1, \ldots, n - p + 1$ are incongruent modulo p. One of them, namely, $n - r$ is congruent to 0. Now,

$$\binom{n}{p} = \frac{n(n-1)\cdots(n-p+1)}{p\,!} = \frac{(n-r)P}{p\,!},$$

where P is the product of the factors $n - i$ for $i \neq r$, which are congruent to $1, 2, \ldots, p - 1$ in some order. Hence, $P \equiv (p-1)!$ $\pmod p$. Since $n - r = mp$, we get

$$p\,!\binom{n}{p} = mp\,\{Np + (p-1)\,!\},$$

$$(p-1)\,!\left\{\binom{n}{p} - m\right\} = Nmp, \qquad (*)$$

for some integer N. But p is relatively prime to $(p-1)!$. Thus,

$$\binom{n}{p} - m \equiv 0 \pmod{p},$$

which proves the first result.

For the second part, suppose that $\binom{n}{p} = kp^a$ and $m = k_1 p^b$, where k, k_1 are not divisible by p, and $a, b \geq 0$. Now, $(p-1)! \equiv -1 \pmod{p}$. If $(p-1)! = N_1 p - 1$, we can rewrite (*) as

$$(N_1 p - 1)(kp^a - k_1 p^b) = N k_1 p^{b+1}.$$

If $a > b \geq 0$, dividing by p^b, we get

$$(N_1 p - 1)\left(kp^{a-b} - k_1\right) = N k_1 p,$$

which implies that $k_1 \equiv 0 \pmod{p}$. If $b > a \geq 0$, dividing by p^a we get,

$$(N_1 p - 1)\left(k - k_1 p^{b-a}\right) = N k_1 p^{b-a}.$$

which forces $k_1 \equiv 0 \pmod{p}$. Both yield contradictions. Hence, $a = b$.

23. Suppose that a, b and n are positive integers and n divides $a^n - b^n$. Prove that n divides $\dfrac{a^n - b^n}{a - b}$.

Solution Suppose that n is a product of k primes, not necessarily distinct. We shall prove the result by induction on k, the number of prime divisors of n.

We have $a^n - b^n = N(a-b)$, where

$$N = a^{n-1} + a^{n-2}b + \cdots + ab^{n-2} + b^{n-1}.$$

If $k = 1$, then n is prime; hence n divides N or $a - b$. If $n|(a-b)$, then

$$a \equiv b \pmod{n}, \quad a^{n-i}b^{i-1} \equiv a^{n-1} \pmod{n} \quad \text{for all } i,$$

and so $N \equiv na^{n-1} \pmod{n}$. Thus, in either case, $N \equiv 0 \pmod{n}$, that is, $N = (a^n - b^n)/(a-b)$ is divisible by n.

Assume that the result is true when n is a product of $k-1$ primes. Let $n = mp$, where p is prime and m is a product of $k-1$ primes. Suppose that $a^n - b^n = N_1 n$ for some integer N_1, and let $x = a^p, y = b^p$. Since $a^n - b^n = x^m - y^m$ is divisible by m, our assumption implies that $(x^m - y^m)/(x - y) = N_2 m$ for some integer N_2. Thus,

$$N_1 mp = N_1 n = a^n - b^n = N_2 m(x - y) = N_2 m(a^p - b^p).$$

This gives $N_1 p = N_2(a^p - b^p)$, and so p divides N_2 or $a^p - b^p$. If p divides $a^p - b^p$, then it divides $(a^p - b^p)/(a - b) = N_3$, say, by the case $k = 1$. In either case, $N_2 N_3$ is divisible by p. Hence,

$$a^n - b^n = N_2 m(a^p - b^p) = N_2 N_3 (a - b)m.$$

Since $m = n/p$ and $(N_2 N_3)/p$ is an integer, we see that $a^n - b^n$ is divisible by $n(a - b)$. This completes the proof by induction.

24. Let a, b and x_0 be positive integers, and define

$$x_n = ax_{n-1} + b \quad \text{for } n = 1, 2, 3, \ldots.$$

Prove that not all the x_n can be primes.

Solution First suppose that $a = 1$. Then $x_n - x_{n-1} = b$, and hence $x_n = x_1 + (n - 1)b$ for $n \geq 1$. If $n = x_1 + 1$, then $x_n = x_1(b + 1)$, which is composite since $x_1 = x_0 + b$ and $b + 1$ are greater than 1.

Next let $a > 1$. Since $\dfrac{x_n}{a^n} = \dfrac{x_{n-1}}{a^{n-1}} + \dfrac{b}{a^n}$, we get

$$\frac{x_n}{a^n} - \frac{x_1}{a} = \frac{b}{a^n} + \frac{b}{a^{n-1}} + \cdots + \frac{b}{a^2},$$

$$x_n - x_1 a^{n-1} = b(1 + a + a^2 + a^{n-2}) = b\frac{a^{n-1} - 1}{a - 1}.$$

If x_1 is composite, there is nothing to prove. Otherwise, let $n = x_1$. Then, n is prime and $n = ax_0 + b > a$. Hence, $a^{n-1} \equiv 1 \pmod{n}$, and

$$(a - 1)x_n = a^{n-1}(a - 1)x_1 + b(a^{n-1} - 1) \equiv 0 \pmod{n}.$$

Since n is relatively prime to $a - 1$, this means that x_n is divisible by n.

25. Let n, r, a denote positive integers. Prove that

$$n^2 \equiv n \pmod{10^a} \Longrightarrow n^r \equiv n \pmod{10^a} \quad \text{for all } r. \qquad (1)$$

Find all values of r such that

$$n^r \equiv n \pmod{10^a} \Longrightarrow n^2 \equiv n \pmod{10^a}. \tag{2}$$

Solution If $n^2 \equiv n \pmod{10^a}$, then $n^r \equiv n^{r-1} \pmod{10^a}$ for all $r \geq 2$. This readily gives (1).

For (2), first let r be odd. If $n = 10^a - 1$, then $n \equiv -1 \pmod{10^a}$, and so $n^r - n \equiv (-1)^r - (-1) \equiv 0 \pmod{10^a}$, while

$$n^2 - n = n(n-1) \equiv (-1)(-2) \pmod{10^a}.$$

Thus (2) is false in this case.

Next consider the case $r \equiv 0, 2, 4$ or $8 \pmod{10}$. If $r = 0$, then $n^r \equiv n \pmod{10^a}$ only when $n \equiv 1 \pmod{10^a}$. So (2) holds if $r = 0$. If $r = 2$, then (2) is trivial. So suppose that $r \geq 4$. We have $n^r - n = n(n^{r-1} - 1) = f(n)g(n)$, where

$$f(n) = n(n-1) = n^2 - n,$$
$$g(n) = n^{r-2} + n^{r-3} + \cdots + n^2 + n + 1.$$

Since r is even, we can write

$$\begin{aligned} g(n) &= 1 + n(1+n) + n^3(1+n) + \cdots + n^{r-3}(1+n) \\ &= 1 + n(n+1)\left\{1 + n^2 + \cdots + n^{r-4}\right\}. \end{aligned}$$

Since $n(n+1)$ is always even, we see that $g(n)$ is odd for all n. This also shows that if $n \equiv 0$ or $-1 \pmod 5$, then $g(n) \equiv 1 \pmod 5$. If $n \equiv 1 \pmod 5$, then $g(n) \equiv r - 1 \not\equiv 0 \pmod 5$. If $n \equiv \pm 2 \pmod 5$, then $n^2 \equiv -1 \pmod 5$, and so $n^r - n \equiv (-1)^{r/2} - n \equiv \pm 1 \pm 2 \pmod 5$. Thus, we see that $g(n) \not\equiv 0 \pmod 5$ for any n. It follows that $g(n)$ is relatively prime to 2 and 5, and hence to 10^a. So

$$f(n)g(n) \equiv 0 \pmod{10^a} \Longrightarrow f(n) \equiv 0 \pmod{10^a}.$$

Thus, (2) holds if $r \equiv 0, 2, 4$ or $8 \pmod{10}$.

Finally, consider the case $r \equiv 6 \pmod{10}$. Since 5^a is relatively prime to 2^a, we can choose a positive integer m such that

$$5^a m \equiv -(5^{a-1} + 1) \pmod{2^a}.$$

Let $n = 5^a m+5^{a-1}+1$. Then, n is divisible by 2^a and $n \equiv 1 \pmod 5$. Hence, $g(n) \equiv r - 1 \equiv 0 \pmod 5$. Let $g(n) = 5k$. Since $n(n-1)$ is divisible by $2^a \cdot 5^{a-1}$, we see that $n^r - n = n(n-1)g(n)$ is divisible by $2^a \cdot 5^a = 10^a$. On the other hand,

$$n(n-1) = \left(5^a m + 5^{a-1} + 1\right)\left(5^a m + 5^{a-1}\right).$$

If $a = 1$, then $n = 5m+1+1$, and so $n(n-1) = (5m+2)(5m+1) \equiv 2 \pmod 5$. If $a > 1$, then

$$n(n-1) \equiv \left(5^{a-1} + 1\right) 5^{a-1} \equiv 5^{a-1} \pmod{5^a},$$

since $2a-2 \geq a$. It follows that $n(n-1)$ is not divisible by 5^a, whether $a = 1$ or $a > 1$. Thus, $n^2 - n \not\equiv 0 \pmod{10^a}$ even though $n^r - n \equiv 0 \pmod{10^a}$. Hence, (2) does not hold in this case.

The values of r for which (2) holds are given by $r \equiv 0, 2, 4$ or $8 \pmod{10}$.

26. Let n, a, d be given integers with $(a, d) = 1$. Prove that there exists an integer m such that $m \equiv a \pmod d$ and $(m, n) = 1$.

Solution If $(a, n) = 1$, we can take $m = a$. So let $(a.n) > 1$. Then there is a prime p dividing a and n. If $p|a$, then p does not divide d since $(a, d) = 1$. Thus, there exists a prime p dividing n but not d. Let S be the set of all primes dividing n but not d. Then, S is non-empty. Let P be the product of the primes in S. Then, $(d, P) = 1$. Hence, by the Chinese remainder theorem, there is an integer m satisfying

$$m \equiv a \pmod d \quad \text{and} \quad m \equiv 1 \pmod P.$$

Let p be a prime dividing n. If $p \in S$, then $p|P$, and hence $m \equiv 1 \pmod p$. Suppose that $p \notin S$, then p divides d also, and so p does not divide a since $(a, d) = 1$. Moreover, if $p|d$, then $m \equiv a \pmod p$, and hence, p does not divide m. Thus, in either case, p does not divide m. This is true for any prime p dividing n. It follows that $(m, n) = 1$. This proves the result.

27. Given that $[x + y] = [x] + [y]$ and $[-x - y] = [-x] + [-y]$, prove that x or y is an integer.

Solution Let $[x] = m$ and $[y] = n$. Assume that neither x nor y is an integer. Then, $m < x < m + 1$, $-m > -x > -m - 1$, and hence, $[-x] = -m-1$. Similarly, $[-y] = -n-1$. The given relations become

$$[x + y] = m + n \quad \text{and} \quad [-x - y] = (-m - 1) + (-n - 1).$$

These imply that

$$x + y < m + n + 1 \quad \text{and} \quad -x - y < -m - n - 1.$$

Thus, our assumption leads to contradictory results.

28. If p is prime, find formulae for the highest power of p that divides

   (a) the product $2 \cdot 4 \cdot 6 \cdots (2n)$ of the first n even integers;

   (b) the product of the first n odd integers.

Solution The exponent of the highest power of a prime p that divides $n!$ is

$$e_p = \left[\frac{n}{p}\right] + \left[\frac{n}{p^2}\right] + \left[\frac{n}{p^3}\right] + \cdots.$$

Let $P_1 = 2 \cdot 4 \cdot 6 \cdots (2n) = 2^n n!$, and P_2 be the product of the first n odd integers. Then, $P_1 P_2 = (2n)!$. The exponent of the highest power of p that divides P_1 is e_p itself if $p \neq 2$; it is $n + e_2$ if $p = 2$.

Let f_p and g_p be the exponents of the highest powers of p that divide P_2 and $(2n)!$, respectively. Then g_p is given by the formula for e_p, with n changed to $2n$. Since $P_1 P_2 = (2n)!$, we see that if $p \neq 2$, then $g_p = e_p + f_p$. Thus,

$$f_p = g_p - e_p = \sum_{i \geq 1} \left[\frac{2n}{p^i}\right] - \sum_{i \geq 1} \left[\frac{n}{p^i}\right]$$

if $p \neq 2$, and $f_2 = 0$ since P_2 is odd.

29. Let p be prime and $n = a_0 + a_1 p + a_2 p^2 + \cdots + a_k p^k$, where a_i are integers with $0 \leq a_i < p$, $a_k \neq 0$. Show that the exponent of the highest power of p that divides $n!$ is

$$\frac{n - (a_0 + a_1 + a_2 + \cdots + a_k)}{p - 1}.$$

Solution Let $1 \leq i \leq k$, and $\dfrac{n}{p^i} = Q_i + N_i$, where

$$Q_i = \frac{a_0 + a_1 p + \cdots a_{i-1} p^{i-1}}{p^i} \leq \frac{(p-1)(1 + p + \cdots p^{i-1})}{p^i} < 1,$$

and $N_i = a_i + a_{i+1}p + \cdots + a_k p^{k-i}$. This shows that $\left[\frac{n}{p^i}\right] = N_i$ for $1 \leq i \leq k$. Hence, the exponent of the highest power of p that divides $n!$ is

$$e = N_1 + N_2 + \cdots + N_k.$$

Using the values of N_i given above, we get

$$e = a_1 + a_2(p+1) + \cdots + a_k(p^{k-1} + p^{k-2} + \cdots + 1).$$

Thus,

$$\begin{aligned} e(p-1) &= a_k(p^k - 1) + \cdots + a_2(p^2 - 1) + a_1(p-1) \\ &= (n - a_0) - (a_1 + a_2 + \cdots + a_k). \end{aligned}$$

This proves the result.

30. Find all integers $n \geq 1$ such that the exponent of the highest power of 5 that divides $n!$ is 100.

Solution Let $n = a_0 + a_1 \cdot 5 + a_2 \cdot 5^2 + \cdots + a_k \cdot 5^k$, where $0 \leq a_i \leq 4$, $a_k \neq 0$. Then, by the previous problem,

$$100(5-1) = a_1(5-1) + a_2(5^2 - 1) + \cdots + a_k(5^k - 1).$$

This first of all implies that $400 \geq a_k(5^k - 1) \geq 5^k - 1$, and so $k \leq 3$. Hence, $400 = a_1(4) + a_2(24) + a_3(124)$. Thus, $400 \geq 124a_3$ and $400 \leq 16 + 96 + 124a_3$. It follows that $a_3 = 3$. So $4a_1 + 24a_2 = 400 - 372 = 28$. This implies that $a_2 = 1$ and $a_1 = 1$. Hence, $n = a_0 + 5 + 5^2 + 3 \cdot 5^3 = 405 + a_0$, where $0 \leq a_0 \leq 4$. Thus, $n = 405$, 406, 407, 408, 409.

Set IV

1. Show that $n^4 + 4$ is divisible by at least two primes for all $n > 1$.

Solution If $n = 2k$ is even, then $n^4 + 4 = 16k^4 + 4 = 4(4k^4 + 1)$, and $4k^4 + 1$ has an odd prime divisor p. So $n^4 + 4$ has at least two prime divisors, namely 2 and p.

Let n be odd. We have

$$n^4 + 4 = (n^2 + 2)^2 - 4n^2 = (n^2 + 2 + 2n)(n^2 + 2 - 2n).$$

Since $1 < n^2+2-2n < n^2+2+2n$ for $n > 1$, we see that n^2-2n+2 and n^2+2n+2 have prime divisors. Let p be a prime dividing both the numbers. Then p divides their difference, that is, $4n$. So $p|4$ or $p|n$. If $p|n$, then $p|(n^2 \pm 2n)$; and hence $p = 2$, in either case. Thus, $2|(n^2 + 2n + 2)$, and so n is even. This contradiction shows that $n^2 + 2n + 2$ and $n^2 - 2n + 2$ must have different prime divisors.

2. Determine all positive integers n such that $n^4 + 4^n$ is divisible by two distinct primes.

Solution If $n = 1, 2, 3$ and 4, then $n^4 + 4^n = 5$, 32, 145 and 512, respectively. Of these, only 145 has two distinct prime divisors. Let $n > 4$.

First suppose that n is even, and let $n = m2^r$, where m is odd and $r \geq 1$. If $m = 1$, then $r > 2$. Then,

$$n^4 + 4^n = m^4 2^{4r} + 4^{m2^r} = 4^{2r}\left(m^4 + 4^N\right),$$

where $N = m2^r - 2r$. If $m > 1$, then $m2^r > 2^r \geq 2r$ for $r \geq 1$. If $m = 1$, then $r > 2$ and hence $m2^r = 2^r > 2r$. Thus $N > 0$ in any case. This implies that $m^4 + 4^N$ has an odd prime divisor p. So $n^4 + 4^n$ has two prime divisors, 2 and p.

Next suppose that n is odd. Then,

$$n^4 + 4^n = (n^2 + 2^n)^2 - 2n^2 2^n = (n^2 + 2^n)^2 - a^2,$$

where $a = n2^{(n+1)/2}$. Since $(n+1)/2$ is an integer, we see that a is an integer, and so $n^4 + 4^n$ is the product of two positive integers, namely, $n^2 + 2^n - a$ and $n^2 + 2^n + a$. These two factors are odd since a is even. Now,

$$n^2 + 2^n - a \geq 2n2^{n/2} - a = (\sqrt{2} - 1)a > 1,$$

since $a > n > 4$. It follows that these two factors have odd prime divisors.

Suppose that d is a common divisor of $n^2 + 2^n + a$ and $n^2 + 2^n - a$. Then d divides their difference, namely $2a$. Since d is odd, this implies that d divides $a = n2^{(n+1)/2}$ and so $d|n$. Hence, we get

$$d|n^2, \quad d|(n^2 + a), \quad d|2^n,$$

successively. Thus, $d = 1$. It follows that the two factors have distinct odd prime divisors.

We conclude that if $n = 3$ or $n > 4$, then $n^4 + 4^n$ have at least two distinct prime divisors.

3. Find the remainder obtained when the number $N = 1^1 + 2^2 + 3^3 + \cdots + 100^{100}$ is divided by 16.

Solution We have $(2n+1)^2 = 4n(n+1)+1 = 8m+1$ for some m. So

$$(2n+1)^4 = (8m+1)^2 \equiv 1 \pmod{16}.$$

Thus, $(4n+1)^{4n+1} \equiv 4n+1 \pmod{16}$, and

$$(4n+3)^{4n+3} \equiv (4n+3)^3 \equiv 3(4n)3^2 + 3^3 \equiv 12n + 11 \pmod{16}.$$

Hence, $(4n+1)^{4n+1} + (4n+3)^{4n+3} \equiv 12 \pmod{16}$. Adding,

$$\sum_{n=0}^{24} \left[(4n+1)^{4n+1} + (4n+3)^{4n+3}\right] \equiv 25 \times 12 \pmod{16}.$$

Since $(2n)^{2n} \equiv 0 \pmod{16}$ for $n \geq 2$, we see that

$$N \equiv 300 + 2^2 \equiv 0 \pmod{16}.$$

So the remainder is 0.

4. If p is a prime number, prove that $2^p + 3^p$ is never a perfect power.

Solution Let $2^p + 3^p = m^k$ for some $k \geq 2$. If $p = 2$, then $13 = m^k$, and hence $m = 13$, $k = 1$. Let p be odd. Then, $2^p + 3^p$ is divisible by 5. Thus, $5|m^k$, $5|m$ and hence $5^k|m^k$. Suppose that $k \geq 2$. Then $(2^p + 3^p)/5$ is divisible by 5. But

$$\begin{aligned} \frac{2^p + 3^p}{5} &= 2^{p-1} - 2^{p-2} \cdot 3 + \cdots - 2 \cdot 3^{p-2} + 3^{p-1} \\ &\equiv 2^{p-1} + 2^{p-2}2 + \cdots + 2^{p-1} \pmod{5}. \end{aligned}$$

The last sum is $p2^{p-1}$, and hence, $(2^p + 3^p)/5 \equiv p2^{p-1} \pmod{5}$. If this is to be divisible by 5, then $5|p$ and so $p = 5$. In that case, $2^p + 3^p = 275 = 25 \cdot 11$. This contradiction shows that $k = 1$.

5. Find all integers n for which $n^2 + n + 41$ is a perfect square.

Solution Let $n^2 + n + 41 = m^2$. Then

$$4m^2 = 4n^2 + 4n + 164 = (2n+1)^2 + 163,$$

and hence $(2m+2n+1)(2m-2n-1) = 163$. Since 163 is a prime number, we must have

$$2m+2n+1 = \pm 1 \quad \text{or} \quad \pm 163,$$

and correspondingly,

$$2m-2n-1 = \pm 163 \quad \text{or} \quad \pm 1.$$

Subtracting, we get $4n+2 = -162$ or 162. Hence $n = -41$ or 40.

6. What is the least number which contains only 1's in its decimal representation and which is divisible by $33\ldots3$, where the digit 3 is repeated 100 times.

Solution Let $R_k = 111\cdots1$, where 1 is repeated k times and $m = R_{100}$. It is required to choose k so that R_k is divisible by $3m$. Let $R_k = N(3m)$. Then, $10^k - 1 = 9R_k = 27Nm$. If $k = 100q + r, 0 \le r < 100$, then $10^k = \left(10^{100}\right)^q \cdot 10^r$. Since $9m = 10^{100} - 1$, we get

$$10^k = (9m+1)^q \cdot 10^r \equiv 10^r \pmod{9m}.$$

Hence, $27Nm \equiv 10^r - 1 \pmod{9m}$. But $10^r - 1 < 10^{100} - 1 = 9m$, which implies that $r = 0$. Since $R_k \equiv k \pmod 3$, we see that k is divisible by 100 and 3. So the least value of k is 300.

7. Which of the integers $101, 10101, 1010101, \ldots$, with alternating 0's and 1's, beginning and ending with 1, can be primes?

Solution Let $N_2 = 101$, $N_3 = 10101$, and in general,

$$N_k = 1010\cdots01 = 1 + 100 + \cdots + 100^{k-1} = \frac{100^k - 1}{100 - 1}.$$

If k is odd and $k > 1$, then

$$N_k = \left(\frac{10^k - 1}{10 - 1}\right)\left(\frac{10^k + 1}{10 + 1}\right).$$

This is composite since $10^k + 1$ is divisible by $10 + 1$ if k is odd. If k is even and $k = 2r$, then

$$N_k = \frac{100^{2r} - 1}{99} = \left(\frac{100^r - 1}{99}\right)(100^r + 1).$$

This is composite if $r > 1$. Thus, only one number in the list, namely $N_2 = 101$, is prime; the remaining are composite.

8. Prove that $2222^{5555} + 5555^{2222}$ is divisible by 7.

Solution We have $2222 \equiv 3 \pmod 7$ and $5555 \equiv 4 \pmod 7$. Hence,

$$2222^{5555} + 5555^{2222} \equiv 3^{5555} + 4^{2222} \pmod 7.$$

By Fermat's theorem, we also have $3^6 \equiv 1 \pmod 7$ and $4^6 \equiv 1 \pmod 7$. Now, $2222 \equiv 2 \pmod 6$ and $5555 \equiv 5 \pmod 6$, that is, $2222 = 6m + 2$ and $5555 = 6n + 5$. It follows that

$$2222^{5555} + 5555^{2222} \equiv 3^{6n+5} + 4^{6m+2} \equiv 3^5 + 4^2 \equiv 0 \pmod 7.$$

9. Prove that a product of n positive integers forming an arithmetic progression whose common difference is relatively prime to $n!$ is divisible by $n!$.

Solution Let $P = a(a + d)(a + 2d) \cdots (a + (n - 1)d)$, where d is relatively prime to $n!$. We can find an integer k such that $kd \equiv 1 \pmod{n!}$, and so

$$k(a + rd) \equiv ka + r \pmod{n!} \quad \text{for } r = 0, 1, \ldots, n - 1.$$

Multiplying we get

$$k^n P \equiv (ka)(ka + 1) \cdots (ka + n - 1) \pmod{n!}.$$

Since the right-hand side is the product of n consecutive integers, it is divisible by $n!$. This gives

$$k^n P \equiv 0 \pmod{n!}, \quad P \equiv k^n d^n P \equiv 0 \pmod{n!},$$

since $kd \equiv 1 \pmod{n!}$.

10. Find all positive integers which can be expressed as the sum of three distinct composite numbers.

Solution If n is the sum of three distinct composite numbers, then $n \geq 4 + 6 + 8 = 18$. Let $n \geq 18$. If n is even, then $n - 10$ is an even number greater than 6, and $n = 4 + 6 + (n - 10)$. If n is odd, then $n - 13$ is an even number greater than 4, and $n = 4 + 9 + (n - 13)$. So the desired positive integers are all $n \geq 18$.

11. Find all natural numbers n such that $[n^2/5]$ is a prime.

Solution For $n = 1, 2, \ldots, 9$, the values of $[n^2/5]$ are 0, 0, 1, 3, 5, 7, 9, 12, 16 respectively. Of these, only 3, 5 and 7 are prime. These correspond to $n = 4$, 5 and 6. Let $n = 5k + r$, where $k \geq 2$ and $0 \leq r \leq 4$. Then

$$\left[\frac{n^2}{5}\right] = \left[5k^2 + 2kr + \frac{r^2}{5}\right] = \begin{cases} 5k^2 + 2kr & \text{for } r = 0, 1, 2 \\ 5k^2 + 6k + 1 & \text{for } r = 3 \\ 5k^2 + 8k + 3 & \text{for } r = 4. \end{cases}$$

Since $5k^2 + 2kr = k(5k + 2r)$, $5k^2 + 6k + 1 = (k + 1)(5k + 1)$ and $5k^2 + 8k + 3 = (k + 1)(5k + 3)$, we see that these are composite. Hence the desired values of n are 4, 5 and 6.

12. Let N be an even number not divisible by 10. What digit is in the ten's place of the number N^{20}?

Solution Let $N = 2n$. Then n is relatively prime to 25. Since $\phi(25) = 25(1 - 1/5) = 20$, we have $n^{20} \equiv 1 \pmod{25}$, by Euler–Fermat theorem. Hence,

$$N^{20} = 2^{20}(25k + 1) \equiv 2^{20} \pmod{100}.$$

Since $2^{10} = 1024 \equiv 24 \pmod{100}$, we get $2^{20} \equiv 576 \equiv 76 \pmod{100}$. Thus, $N \equiv 76 \pmod{100}$, and the last two digits in N^{20} are 7, 6. Hence the digit in the ten's place is 7.

13. Find all primes p such that $\dfrac{2^{p-1} - 1}{p}$ is a perfect square.

Solution If $p = 2$, then $(2^{p-1} - 1)/p$ is not an integer. If $p = 3$, 5 and 7, then $(2^{p-1} - 1)/p = 1$, 3 and 9, respectively. Thus, $(2^{p-1} - 1)/p$ is a perfect square when $p = 3$ and 7.

Let p be a prime greater than 7, and $p = 2n + 1$. Then $n > 3$, and

$$2^{p-1} - 1 = 2^{2n} - 1 = (2^n + 1)(2^n - 1).$$

Hence, p divides $2^n + 1$ or $2^n - 1$. However, $2^n + 1$ and $2^n - 1$ are coprime as they are odd. If $p | 2^n + 1$, then $2^n - 1$ and $(2^n + 1)/p$ are coprime integers and so they must be perfect squares. But $2^n - 1 \equiv -1 \pmod{4}$, and any perfect square is congruent to 0 or 1 modulo 4.

In the other case, $p | 2^n - 1$. Then, $2^n + 1$ is a perfect square. If $2^n + 1 = a^2$, then $2^n = (a - 1)(a + 1)$, and hence

$$a - 1 = 2^r, \quad a + 1 = 2^k, \quad 2 = 2^k - 2^r = 2^r(2^{k-r} - 1).$$

This implies that $r = 1$, $k = 2$, and so $n = 3$, $p = 7$. Thus, we see that there is no prime $p > 7$ for which $(2^{p-1} - 1)/p$ is a perfect square.

14. Find all natural numbers n such that $n^3 - 18n^2 + 115n - 391$ is cube of some natural number m.

Solution Let m, n be natural numbers such that $n^3 - 18n^2 + 115n - 391 = m^3$. This can be written as

$$(n-6)^3 + 7(n-25) = m^3, \quad k^3 + 7(k-19) = m^3,$$

where $k = n - 6$. If $k = 19$, then $m = k$. If $k > 19$, then $m > k$ and so $m \geq k + 1$,

$$k^3 + 7(k-19) = m^3 \geq k^3 + 3k^2 + 3k + 1.$$

This shows that $3k^2 - 4k + 134 \leq 0$. This is not possible since

$$3k^2 - 4k + 134 = k^2 + 2(k-1)^2 + 132.$$

Next suppose that $k < 19$. Then, $m < k$, and hence, $m \leq k - 1$,

$$k^3 + 7(k-19) = m^3 \leq k^3 - 3k^2 + 3k - 1,$$

which holds only if

$$3k^2 + 4k - 132 \leq 0, \quad (k-6)(3k+22) \leq 0.$$

So $k \leq 6$. Again, $k^3 + 7k = m^3 + 133 \geq 134$, which implies that $k \geq 5$. Hence $k = 5$ or 6.

Thus, the values of n, for which the given expression is the cube of a positive integer m, are given by $n = k + 6$, where $k = 19$, 5 or 6. The corresponding values of m are given by $m^3 = k^3 + 7(k-19)$, that is,

$$m^3 = 19^3, \quad 5^3 + 7(-14) = 3^3, \quad 6^3 + 7(-13) = 5^3.$$

15. Show that $x^3 + y^3 + z^3 + t^3 = 1999$ has infinitely many integer solutions.

Solution We have $(a+10)^3 - (a-10)^3 = 60a^2 + 2000$. If $a = 60n^3$, then $60a^2 = (60n^2)^3$ and hence,

$$\left(60n^3 + 10\right)^3 - \left(60n^3 - 10\right)^3 - (60n^2)^3 - 1 = 1999.$$

This is true for all integers n.

16. The arithmetic mean of two or more distinct primes is 27. Find the greatest of such primes.

Solution Let $p_1 < p_2 < \cdots < p_n$ and $p_1 + p_2 + \cdots + p_n = 27n$. Since $27n \equiv n \pmod 2$ and $p_2 + \cdots + p_n \equiv (n-1) \pmod 2$, we see that p_1 is odd. If $p_k < 27 < p_{k+1}$, then $p_1, \ldots, p_k$ are primes in $\{3, 5, \ldots 23\}$. Since $\sum_{i=1}^{n}(p_i - 27) = 0$, we get

$$p_n - 27 \leq \sum_{i=1}^{k}(27 - p_i) \leq (27-3) + \cdots + (27-23) = 118.$$

Thus, $p_n \leq 139$. Since

$$\frac{1}{9}(3+5+7+11+13+17+19+29+139) = 27,$$

the answer is 139.

17. Given m integers $a_1, a_2, \ldots, a_m$, show that there exist k, r with $0 \leq k < r \leq m$ such that $a_{k+1} + a_{k+2} + \cdots + a_r$ is divisible by m.

Solution Let $b_n = \sum_{i=1}^{n} a_i$, $1 \leq n \leq m$. If some b_n is divisible by m, we can take $k = 0$, $r = n$. So suppose that none of $b_1, \ldots, b_m$ is divisible by m. Then they are congruent modulo m to $1, 2, \ldots, m-1$ in some order, and hence two of them are congruent. If $b_k \equiv b_r \pmod m$, then $a_{k+1} + \cdots + a_r$ is divisible by m.

18. Suppose that 101 numbers are chosen from the set $S = \{1, 2, 3, \ldots, 200\}$. Show that two of the chosen numbers are such that one is divisible by the other.

Solution Each integer in S can be expressed as $2^k\, m$, where m is odd. The 'odd part' m can be any of the 100 numbers $1, 3, \ldots, 199$. So two of them have the same odd part. If a and b have the same odd part, then $a = 2^k\, m$ and $b = 2^r\, m$. Hence $a|b$ or $b|a$.

19. Show that among any 8 composite numbers selected from the first 360 natural numbers, there will always be two which are not relatively prime.

Solution If n is composite, then it is divisible by a prime number not exceeding $\sqrt{n}$. So each of these 8 composite numbers must be divisible by one of the seven primes 2, 3, 5, 7, 11, 13 or 17. Hence two of them must have a common prime divisor.

20. Let p be an odd prime. Show that the numerator of $\sum_{k=1}^{p-1} \frac{1}{k}$ is divisible by p.

Solution Let $a_k = \frac{(p-1)!}{k}$ for $1 \le k \le p-1$. Then, $a_k + a_{p-k} = (p-1)!\,\frac{p}{k(p-k)}$. Obviously this is an integer divisible by p. Hence if the given sum is equal to a/b, then

$$(p-1)!\,\frac{a}{b} = \sum_{k=1}^{(p-1)/2} (a_k + a_{p-k}) = pN,$$

say. Thus, $a(p-1)!$ is divisible by p. Since $(p-1)!$ is relatively prime to p, we get $p|a$.

21. Let m and n be relatively prime integers with $m > n > 1$. Prove that $\frac{\log_{10} m}{\log_{10} n}$ cannot be a rational number.

Solution Assume that $\frac{\log_{10} m}{\log_{10} n} = \frac{r}{s}$, where r, s are positive integers. Then $s \log_{10} m = r \log_{10} n$ and hence, $m^s = n^r$. Since $n > 1$, it has a prime divisor p. Then p divides m. This is not possible since m is relatively prime to n.

22. Let a, b be positive integers such that $\frac{a+1}{b} + \frac{b+1}{a}$ is also an integer. Prove that $\gcd(a, b) \le \sqrt{a+b}$.

Solution Let $d = \gcd(a, b)$ so that $a = md$ and $b = nd$, where $(m, n) = 1$. If $(a+1)/b + (b+1)/a = c$, then $a^2 + a + b^2 + b = abc$, and so

$$d^2(m^2 + n^2) + d(m+n) = d^2 mnc.$$

This shows that $d|(m+n)$ and hence, $d \le m+n$. Thus, $d^2 \le d(m+n) = a+b$.

23. Determine all real numbers x satisfying the equation

$$[x] + [2x] + [4x] + [8x] + [16x] + [32x] = 12345.$$

Solution If $m = [y]$, then $m \le y < m+1$. Thus,

$$12345 \le x + 2x + 4x + 8x + 16x + 32x < 12345 + 6,$$

that is, $12345 \le 63x < 12351$. This shows that $195 < x < 197$, and hence $x = 195 + t$, where $0 \le t < 2$. Note that $[n + y] = n + [y]$. So the given equation becomes

$$195 \times 63 + [t] + [2t] + [4t] + [8t] + [16t] + [32t] = 12345.$$

Let $m = [t] + [2t] + [4t] + [8t] + [16t] + [32t]$. Then, $m = 60$. If $0 \le t < 1$, then $0 \le [kt] \le k - 1$ and hence,

$$0 \le m \le 0 + 1 + 3 + 7 + 15 + 31 = 57.$$

If $1 \le t < 2$, then $k \le [kt]$ and hence,

$$1 + 2 + 4 + 8 + 16 + 32 \le m.$$

In either case, we cannot have $m = 60$. The given equation has no solution.

24. Let $s(N)$ be the sum of digits in the decimal representation of N. Prove that there are infinitely many values of n for which $s(2^n) > s(2^{n+1})$.

Solution Assume, on the contrary, that $s(2^n) \le s(2^{n+1})$ for all sufficiently large n. First observe that $s(m) \equiv m \pmod 9$, and also that $2^{6n} = 64^n \equiv 1 \pmod 9$. It follows that

$$s(2^{6n+i}) \equiv 2^{6n+i} \equiv 2^i \pmod 9,$$

and hence, $s(2^{6n+i+1}) - s(2^{6n+i}) \equiv 2^{i+1} - 2^i \pmod 9$. But $2^{i+1} - 2^i = 2^i$. Let $2^i \equiv r_i \pmod 9$, where $0 \le r_i \le 8$. Then

$$s(2^{6n+i+1}) - s(2^{6n+i}) = 9N + r_i,$$

for some integer N. If $N \le -1$, then $9N + r_i < 0$, and so

$$s(2^{6n+i+1}) - s(2^{6n+i}) < 0.$$

By our assumption, we must have $s(2^{m+1}) - s(2^m) \ge 0$ for all large m. Consequently, if n is large, then $N \ge 0$, and hence,

$$s(2^{6n+i+1}) - s(2^{6n+i}) \ge r_i.$$

This holds for $0 \le i \le 5$. But the values of r_i for $i = 0, 1, \ldots, 5$ are 1, 2, 4, 8, 7, 5, respectively. Adding these inequalities, we get

$$s(2^{6n+6}) - s(2^{6n}) \ge 1 + 2 + 4 + 8 + 7 + 5 = 27.$$

This holds for all large n, say for $n \geq n_0$. Repeated application of this with $n = n_0, n_0 + 1, \ldots$ shows that

$$s(2^{6n}) - s(2^{6n_0}) \geq 27(n - n_0), \quad \text{for all } n > n_0.$$

Suppose that 2^{6n} has k digits. Then, $10^{k-1} \leq 2^{6n} < 10^{2n}$. So $k - 1 < 2n$, and hence $s(2^{6n}) \leq 9k \leq 18n$. Hence,

$$27(n - n_0) < s(2^{6n}) \leq 18n,$$

which gives $9n < 27n_0$. This cannot hold for large n. The contradiction proves the result.

25. Determine the remainder when 3^{2^n} is divided by 2^{n+3}.

Solution We shall prove that the required remainder is $1 + 2^{n+2}$. If n is 1 and 2, the remainders are 9 and 17. This agrees with the assertion when $n = 1, 2$. Assume that it is true for some n. Then $3^{2^n} = 1 + 2^{n+2} + N\,2^{n+3}$ for some integer N, and

$$\begin{aligned} 3^{2^{n+1}} &= \left(3^{2^n}\right)^2 = 1 + 2\left(2^{n+2} + N\,2^{n+3}\right) + \left(2^{n+2} + N\,2^{n+3}\right)^2 \\ &= 1 + 2^{n+3} + N_1\,2^{n+4}, \end{aligned}$$

for some N_1. This proves our assertion by induction.

26. Prove that, for each n, there exists a positive integer A_n divisible by 2^n and whose decimal representation contains exactly n digits each of which is 1 or 2.

Solution There are exactly 2^n numbers having n digits each of which is 1 or 2. We shall prove that any two such numbers are incongruent modulo 2^n. Then one of them must be congruent to 0 modulo 2^n.

Since $1, 2$ are incongruent modulo 2, and $11, 12, 21, 22$ are incongruent modulo 4, our claim holds when $n = 1, 2$. Assume that our claim holds for some n. Let N_1 and N_2 be two such numbers having $n + 1$ digits. Then $N_1 = 10m_1 + d_1$ and $N_2 = 10m_2 + d_2$, where d_1, d_2 are 1 or 2. If $d_1 \neq d_2$, then $N_1 - N_2$ is odd. So if $N_1 \equiv N_2 \pmod{2^{n+1}}$, then $d_1 = d_2$, and hence $N_1 - N_2 = 10(m_1 - m_2)$. Thus,

$$N_1 - N_2 \equiv 0 \pmod{2^{n+1}} \implies \quad m_1 - m_2 \equiv 0 \pmod{2^n}.$$

But m_1 and m_2 are positive integers whose decimal representations contain exactly n digits each of which is 1 or 2. This is possible only when $m_1 = m_2$, that is, when $N_1 = N_2$ and thus we prove our claim by induction.

27. Find all integers of the form 2^n such that after deleting the first digit (from the left), we again get a power of 2.

Solution Suppose that 2^n has $k+1$ digits of which the first is a, and $k+1 \geq 2$. If $k=1$, then 2^n has two digits, and so $2^n = 16$, 32 or 64. Of these, only 32 and 64 have the desired property.

Next, consider the case $k \geq 2$. Suppose that, after deleting the first digit a, we get 2^m. Then, $2^n = a \cdot 10^k + 2^m$. Since 2^m has k digits and $k \geq 2$, we have

$$2^m \geq 10^{k-1} > 8^{k-1},$$

and hence, $m > 3(k-1)$. Further,

$$a \cdot 5^k = 2^{m-k}\left(2^{n-m} - 1\right).$$

Since $m - k \geq 2k - 2$, this shows that a is divisible by 2^{2k-2}. If $k = 2$, then $2k - 2 \geq 2$ and hence, a is divisible by 2^2. If $k \geq 3$, then $2k - 2 \geq 4$, and a is divisible by 2^4. But a cannot be divisible by 2^4. So we must have $k = 2$ and $a = 4$ or 8. If $k = 2$, then 2^n is a three-digit number. There is no three-digit power of 2 whose first digit is 4 or 8. So 32 and 64 are the only powers of 2 with the desired property.

28. Let $\{a_n\}$ be a strictly increasing sequence of positive integers such that

   (i) $a_{2n} = a_n + n$ for all $n \geq 1$, and

   (ii) if a_n is prime, then n is prime.

Prove that $a_n = n$ for all n.

Solution By (i), $a_2 = a_1 + 1$ and $a_4 = a_2 + 2 = a_1 + 3$. Since $a_1 + 1 = a_2 < a_3 < a_4 = a_1 + 3$, we get $a_3 = a_1 + 2$.

Assume that $a_n = a_1 + n - 1$ for $1 \leq n \leq 2^k$. This is true when $k = 2$. For $2 < m \leq 2^k$, we have

$$a_{2m} = a_m + m = a_1 + 2m - 1,$$

$$a_{2m-2} = a_{m-1} + m - 1 = a_1 + 2m - 3.$$

Since a_{2m-1} lies strictly between a_{2m} and a_{2m-2}, this implies that $a_{2m-1} = a_1 + 2m - 2$. Thus, $a_n = a_1 + n - 1$ when $n = 2m$ and $n = 2m - 1$, where $2 < m \leq 2^k$. The validity of this for $2^{k-1} < m \leq 2^k$ shows that $a_n = a_1 + n - 1$ for $2^k < n \leq 2^{k+1}$.

Thus, we have proved by induction that $a_n = a_1 + n - 1$ for all $n \geq 1$. Let $a_1 = c$, and suppose that $c > 1$. If p is any prime, then $a_{p-c+1} = a_1 + p - c = p$. Hence, $p - c + 1$ is prime by (ii). If $c > 1$, choose c consecutive composite numbers $N+1, N+2, \ldots, N+c$. Let p be the smallest prime greater than $N + c$. Since $p - c + 1 < p$, we see that $p-c+1$ is a prime less than $N+1$. But $p-c+1 > N+1$ if $p > N+c$. This contradiction shows that $c = 1$. Hence $a_n = a_1 + n - 1 = n$ for all $n \geq 1$.

29. Find the least lcm of 20 natural numbers, not necessarily distinct, whose sum is 801.

Solution At least one of the 20 numbers must exceed 40; otherwise, their sum would not exceed 20×40 which is less than 801. Hence, the lcm must be more than 40. If the lcm is 41, then the numbers can be 1 or 41 only. If n of them are equal to 41, then the sum is $41n + (20 - n) = 40n + 20 \neq 801$. If 19 of the numbers are equal to 42, and the remaining is 3, then the sum is 801 and lcm is 42. Hence, the answer is 42.

30. Find all pairs of natural numbers whose sum is 5432 and least common multiple is 223 020.

Solution Let $d = \gcd(a, b)$, $a_1 = a/d$ and $b_1 = b/d$. Then, $\gcd(a_1, b_1) = 1$, and $\text{lcm}[a, b] = a_1 b_1 d$. So,

$$\begin{aligned} \gcd(\text{lcm}[a, b], a + b) &= \gcd(a_1 b_1 d, (a_1 + b_1)d) \\ &= d \gcd(a_1 b_1, a_1 + b_1). \qquad (*) \end{aligned}$$

If $a_1 x + b_1 y = 1$, then $a_1(x - y) + (a_1 + b_1)y = 1$ which shows that $a_1 + b_1$ is relatively prime to a_1. Similarly, it is relatively prime to b_1. Hence, $a_1 + b_1$ is relatively prime to $a_1 b_1$. It follows from (*) that

$$\gcd(\text{lcm}[a, b], a + b) = d.$$

Here $\text{lcm}[a, b] = 223\,020$ and $a + b = 5432$. Since $223\,020 = 2^2 \cdot 3^3 \cdot 5 \cdot 7 \cdot 59$ and $5432 = 2^3 \cdot 7 \cdot 97$, the gcd of these two is $2^2 \times 7 = 28$. Thus, $d = 28$ and $ab = 28 \times 223\,020$. Let $a = 28x$ and $b = 28y$. Then,

$$x + y = 2 \times 97, \quad xy = 3^3 \times 5 \times 59.$$

This gives $x = 59$, $y = 135$, or $x = 135$, $y = 59$. Hence $a = 1652$, $b = 3780$.

Set V

1. Find all primes p and q such that $p^2 + 7pq + q^2$ is the square of an integer.

   Solution If $p = q$, then $p^2 + 7pq + q^2 = (3p)^2$. So we can have $p = q$, where p is any prime.

   Suppose that p and q are distinct primes and $a^2 = p^2 + 7pq + q^2$, where a is a positive integer. Then,

   $$a^2 - (p+q)^2 = 5pq, \quad (a+p+q)(a-p-q) = 5pq.$$

   Thus, $a + p + q$ and $a - p - q$ are positive divisors of $5pq$. But the possible positive divisors of $5pq$ are $1, 5, p, q, 5p, 5q, pq$ and $5pq$. Since

   $$a+p+q = a-p-q+2(p+q) > 2(p+q),$$

   and $p + q > p, p + q > q, p + q \geq 5$, we see that

   $$a+p+q = 5p,\ 5q,\ pq \quad \text{or} \quad 5pq.$$

   Correspondingly,

   $$a-p-q = q,\ p,\ 5 \quad \text{or} \quad 1.$$

   Subtracting,

   $$2p+2q = 5p-q,\ 5q-p,\ pq-5 \quad \text{or} \quad 5pq-1. \qquad (*)$$

   The first two possibilities give $3q = 3p$ which is ruled out. The third possibility implies that $(p-2)(q-2) = 9$. Since $p \neq q$, this is possible only when $p - 2 = 1, 9$, and $q - 2 = 9, 1$. Thus, $p = 3, q = 11$, or $q = 3, p = 11$. In these cases, $p^2 + 7pq + q^2 = 19^2$. In the last possibility in (*) we get

   $$10p+10q = 25pq-5, \quad (5p-2)(5q-2) = 9.$$

   This does not give primes p, q. Thus, if p and q are distinct primes, then $(p, q) = (3, 11)$ or $(11, 3)$.

2. If a, b, c, d are positive integers such that $a^5 = b^4$, $c^3 = d^2$, and $c - a = 19$, find $d - b$.

Solution Let $a = \prod_{i=1}^{k} p_i^{a_i}$, $b = \prod_{i=1}^{k} p_i^{b_i}$, where the p_i are distinct primes, $a_i \geq 0$, $b_i \geq 0$. Since $a^5 = b^4$, we see that $\prod p_i^{5a_i} = \prod p_i^{4b_i}$. This implies that $5a_i = 4b_i$, and hence, $4|a_i$, for each i. Let $a_i = 4s_i$, and $m = \prod_i p_i^{s_i}$. Then, $a = m^4$. Similarly, since $c^3 = d^2$, we get $c = n^2$ for some n. Hence,

$$19 = n^2 - m^4 = (n - m^2)(n + m^2).$$

Since $n+m^2 > 1$, we get $n+m^2 = 19$ and $n-m^2 = 1$. Thus, $n = 10$, $m = 3$, and

$$a = 3^4, \quad c = 10^2, \quad d = 10^3 = 1000, \quad b = 3^5 = 243.$$

Hence, $d - b = 757$.

3. Find the largest possible value of k for which 3^{11} is expressible as the sum of k consecutive positive integers.

Solution Suppose that $n + (n + 1) + \cdots + (n + k - 1) = 3^{11}$. Then, $2 \cdot 3^{11} = k(2n + k - 1)$, and hence k divides $2 \cdot 3^{11}$. Thus, $k = 2^r 3^s$, where $r = 0, 1$ and $s = 0, 1, \ldots, 11$. Now, $k \leq 2n + k - 1$ so that $k^2 \leq 2 \cdot 3^{11}$. If $s \geq 6$, then $k^2 \geq 3^{12} > 2 \cdot 3^{11}$. So $s \leq 5$. The greatest value of k is given by $k = 2^r \cdot 3^s$, where $r = 1, s = 5$, that is, $k = 2 \cdot 3^5 = 486$. The corresponding value of n is given by $2n + k - 1 = (2 \cdot 3^{11})/k = 3^6$, that is, $n = 122$.

4. Find the smallest positive integer whose cube ends with 888 in decimal notation.

Solution If the decimal representation of n^3 ends with 888, then $n^3 \equiv 888 \pmod{1000}$. Hence n is even and $(n/2)^3 \equiv 111 \pmod{125}$. So the problem is to find the smallest positive integer $x = n/2$ such that $x^3 \equiv 111 \pmod{125}$.

If x is a solution of this congruence, then $x^3 \equiv 1 \pmod 5$. So $x \equiv 1 \pmod 5$, $x = 5a+1$. Hence, $x^3 \equiv 1+3(5a) \pmod{25}$. Since $x^3 \equiv 11 \pmod{25}$, we get

$$15a + 1 \equiv 11 \pmod{25}, \quad 3a \equiv 2 \pmod 5.$$

Thus, $a \equiv 4 \equiv -1 \pmod 5$, $a = 5b - 1$. Hence, $x = 5(5b - 1) + 1 = 25b - 4$. So $x^3 \equiv -64 + 3(16)(25b) \pmod{125}$. Since $x^3 \equiv 111 \pmod{125}$, we get

$$1200b \equiv 175 \pmod{125}, \quad 75b \equiv 50 \pmod{125}.$$

Thus, we have $3b \equiv 2 \pmod 5$, $b \equiv 4 \pmod 5$, and $b = 5c + 4$. Hence, $x = 25(5c+4) - 4 \equiv 96 \pmod{125}$. The smallest positive value of x is 96; so $n = 192$.

5. Find the remainder obtained when $1997^{99} - 1997$ is divided by $1997^2 + 1998$.

Solution Let $x = 1997$. The problem is to find the remainder when $x^{99} - x$ is divided by $x^2 + x + 1$. Since $x^{99} - 1$ is divisible by $x^3 - 1$ and hence by $x^2 + x + 1$, we see that

$$x^{99} - x = x^{99} - 1 - (x - 1) = q(x)(x^2 + x + 1) - (x - 1),$$

where $q(x)$ is a polynomial with integer coefficients. So the required remainder is $x^2 + x + 1 - (x - 1) = 1997^2 + 2 = 3\,988\,011$.

6. Let n be a positive integer which is not divisible by 2 or 5. Prove that there exists an infinite number of numbers which are divisible by n and whose decimal representations contain only the digit 1.

Solution Let $R_k = 11\cdots 1$, where the digit 1 is repeated k times. Then, $R_k = (10^k - 1)/9$. If n is not divisible by 2 or 5, then 10 is relatively prime to $9n$ and hence, $10^{\phi(9n)} \equiv 1 \pmod{9n}$. This implies that if k is a multiple of $\phi(9n)$, then

$$9R_k = 10^k - 1 \equiv 0 \pmod{9n}, \quad R_k \equiv 0 \pmod n.$$

Thus, $n | R_k$ whenever k is a multiple of $\phi(9n)$.

7. Find all positive integers n such that $n + 9$, $16n + 9$ and $27n + 9$ are all perfect squares.

Solution Suppose that $n + 9 = a^2$, $16n + 9 = b^2$ and $27n + 9 = c^2$, where a, b, c and n are positive integers. Then, $16a^2 - b^2 = 135$, that is, $(4a - b)(4a + b) = 135$. Hence, $4a - b$ is a positive divisor of 135, and $(4a - b)^2 < 135$ as $4a - b < 4a + b$. Since $135 = 27 \cdot 5$, the possible values of $4a - b$ are 1, 3, 9, 5. The corresponding values of $4a + b$ are 135, 45, 15, 27, respectively. It follows that

$$8a = (4a - b) + (4a + b) = 136, 48, 24 \text{ or } 32.$$

Thus, $a = 17, 6, 3$ or 4, and correspondingly, $n = 280, 27, 0$ or 7. But $n > 0$ and $c^2 = 9(3n + 1)$. So $3n + 1$ is a square. This is possible only when $n = 280$. If $n = 280$, then $n + 9 = 17^2$, $16n + 9 = 67^2$ and $27n + 9 = 87^2$.

8. Prove that 2^n is the sum of squares of four positive integers if and only if n is even.

Solution If n is even, then

$$2^n = \left(2^{n/2-1}\right)^2 + \left(2^{n/2-1}\right)^2 + \left(2^{n/2-1}\right)^2 + \left(2^{n/2-1}\right)^2 .$$

Suppose, on the other hand, that n is and 2^n is the sum of four squares. Choose the least such n, and let $2^n = a^2 + b^2 + c^2 + d^2$. Clearly, $n > 3$. Any perfect square is congruent to 0, 1 or 4 modulo 8. If all the numbers a, b, c, d are odd, then $a^2 + b^2 + c^2 + d^2$ is congruent modulo 8 to $1 + 1 + 1 + 1$. But if $n > 2$, then 2^n is divisible by 8. So this is possible only when $n = 2$. If one, two or three of them are odd and the remaining even, then $a^2 + b^2 + c^2 + d^2$ is congruent to $1, 2$ or 3, modulo 8. This shows that if $n > 2$, then a, b, c, d are all even. In that case,

$$2^{n-2} = \left(\frac{a}{2}\right)^2 + \left(\frac{b}{2}\right)^2 + \left(\frac{c}{2}\right)^2 + \left(\frac{d}{2}\right)^2 .$$

Thus, $n - 2$ has the specified property. This contradicts the minimality of n, and completes the proof.

9. If m, n, k are positive integers, prove that

(a) $(mn)\,!$ is divisible by $m\,!\,(n\,!)^m$,

(b) $(mnk)\,!$ is divisible by $m\,!\,(n\,!)^m\,(k\,!)^{mn}$,

(c) $(n^2)!$ is divisible by $(n!)^{n+1}$,

(d) $(n\,!)\,!$ is divisible by $(n\,!)^{(n-1)!}$.

Solution (a) If $m = 1$, then $(mn)\,! = n\,! = (n\,!)^m\, m\,!$. So the result holds when $m = 1$. Similarly, it is true when $n = 1$. Let $m > 1, n > 1$.

We have $\binom{kn}{n} = k\binom{kn-1}{n-1}$ for all $k \geq 1$. Hence,

$$\frac{kn(kn-1)\cdots(kn-n+1)}{n\,!} = ka_k,$$

where a_k is an integer. Put $k = 1, 2, \ldots, m$ in this, and multiply the resulting equalities. The product of the numerators is equal to $(mn)\,!$, and that of the denominators is $(n\,!)^m$. Hence, the product is equal to $(1 \cdot 2 \cdots m)(a_1 \cdot a_2 \cdots a_m)$, a multiple of $m\,!$. This gives (a).

(b) Let $r = mn$. By (a) we see that $r\,! = (mn)\,!$ is divisible by $m\,!\,(n\,!)^m$, and $(mnk)\,! = (rk)\,!$ is divisible by $r\,!\,(k\,!)^r$. Combining these two, we get (b).

(c) Use (a) with $m = n$. Then $mn = n^2$, which is divisible by $(n\,!)^n\,n\,! = (n\,!)^{n+1}$.

(d) Use (a) with $m = (n-1)\,!$. Then $mn = n\,!$, which is divisible by $(n\,!)^m\,m\,!$, and hence by $(n\,!)^m$.

10. Let $a_1, a_2, \ldots, a_r$ be positive integers, and $n = a_1 + a_2 + \cdots + a_r$. Prove that

(a) $N = \dfrac{n\,!}{a_1\,!\,a_2\,!\,\cdots\,a_r\,!}$ is an integer.

(b) If, further, the gcd of $a_1, a_2, \ldots, a_r$ is d, then $N_1 = \dfrac{(n-1)\,!\,d}{a_1\,!\,a_2\,!\,\cdots\,a_r\,!}$ is an integer.

Solution (a) We prove the result by induction on r. If $r = 1$, there is nothing to prove. If $r = 2$, then $N = \binom{n}{a_1}$ is an integer. Let $r > 2$, and assume that the result is true for $r - 1$. If $m = a_1 + \cdots + a_{r-1}$, then $n = m + a_r$ and

$$N = \frac{n\,!}{m\,!\,a_r\,!} \cdot \frac{m\,!}{a_1\,!\cdots a_{r-1}\,!} = \binom{n}{m}\frac{m\,!}{a_1\,!\cdots a_{r-1}\,!}.$$

This is an integer if the result holds for $r - 1$.

(b) If $a_1 > 1$, then $\dfrac{(n-1)\,!}{(a_1-1)\,!\,a_2\,!\,\cdots\,a_r\,!}$ is an integer by (a). Hence, $x_1 = a_1 N_1/d$ is an integer. This is true even when $a_1 = 1$. Similarly, $x_i = a_i N_1/d$ is an integer for each i. There exist integers $y_1, \ldots, y_r$ such that

$$a_1 y_1 + a_2 y_2 + \cdots + a_r y_1 = d.$$

Hence,

$$N_1 = \frac{N_1}{d}\,d = \frac{N_1}{d}\,(a_1 y_1 + \cdots + a_r y_1) = x_1 y_1 + \cdots x_r y_r,$$

which is an integer.

11. The difference between two prime numbers is 100. If we concatenate them (that is, juxtapose one next to another) in one order, we obtain another prime number. Find the two primes and the concatenated prime.

Solution Suppose that p and q are primes such that $p = q + 100$. If $q \equiv 2 \pmod 3$, then $p \equiv 2 + 1 \pmod 3$, which is not possible. If $q \equiv 1 \pmod 3$, then $p \equiv 1 + 1 \pmod 3$. In that case, the sum of the digits in p and q are congruent to 2 and 1, respectively, and hence the sum of the digits in the concatenated number is congruent to $2 + 1$ modulo 3. So the concatenated number is divisible by 3, which is not possible since it is prime. Hence, the only possibility is that $q \equiv 0 \pmod 3$, that is, $q = 3$. Thus, $p = 103$, and the concatenated number is 1033.

12. Find all the primes p and q, and even numbers $n > 2$, satisfying the equation

$$p^n + p^{n-1} + \cdots + 1 = q^2 + q + 1.$$

Solution Since $n > 2$, we have $p^n + p^{n-1} + \cdots + 1 > p^2 + p + 1$. So we cannot have $q = p$. Let $n = 2k$, where $k \geq 2$. Suppose that $q \leq p^k$. Then,

$$q^2 + q + 1 \leq p^{2k} + p^k + 1 < p^{2k} + p^{2k-1} + \cdots + p^k + \cdots + 1.$$

Hence, $q \geq p^k + 1$. The given equation can be written as

$$p(p^{2k-1} + p^{2k-2} + \cdots + 1) = q(q+1),$$

$$q(q+1)(p-1) = p(p^{2k} - 1) = p(p^k - 1)(p^k + 1). \qquad (*)$$

Since $q \neq p$, this implies that q divides $p^k - 1$ or $p^k + 1$. If q divides $p^k - 1$, then $q \leq p^k - 1$, and if q divides $p^k + 1$, then $q \leq p^k + 1$. We found above that $q \geq p^k + 1$. So the only possibility is that $q = p^k + 1$. It follows from (*) that $(q+1)(p-1) = p(q-2)$. Thus,

$$p - q - 1 = -2p, \quad 3p = q + 1 = p^k + 2.$$

This shows, first of all, that $p = 2$, and then $q = 5$, $k = 2$, $n = 4$, successively.

13. Find all pairs (x, y) of integers such that $x^3 + 11^3 = y^3$.

Solution If (x, y) is a solution, then $y^3 > x^3$, and so $y > x$. Further, $11^3 = (y - x)(y^2 + yx + x^2)$. Thus, $y - x$ is a positive divisor of 11^3. This implies that $y - x = 11^r$, where $r = 0, 1, 2$ or 3. Since

$$y^2 + yx + x^2 = \frac{1}{4}(y-x)^2 + \frac{3}{4}(y+x)^2 \geq \frac{1}{4}(y-x)^2,$$

we see that $11^3 \geq \frac{1}{4}(y-x)^3 = \frac{1}{4}(11^{3r})$. Hence, $11^{3r} \leq 4 \cdot 11^3 < 11^4$; $r = 0$ or 1. If $r = 0$, then $y - x = 1$. But x^3, y^3 and 11^3 are congruent modulo 3 to x, y and 11, respectively. Thus, $y - x \equiv 11 \equiv 2 \pmod 3$. So we cannot have $y - x = 1$. Hence, $r = 1$ and $y - x = 11$. The given equation reduces to

$$x^3 + 11^3 = 11^3 + 3(11^2)x + 3(11)x^2 + x^3.$$

This gives $x = 0$ or -11. So there are only two solutions, namely, $x = 0, y = 11$ and $x = -11, y = 0$.

14. Let S denote the set of all 6-tuples (a, b, c, d, e, f) of positive integers such that $a^2 + b^2 + c^2 + d^2 + e^2 = f^2$. Consider the set

$$T = \{abcdef : (a, b, c, d, e, f) \in S\}.$$

Find the greatest common divisor of all members of T.

Solution If none of the five integers a, b, c, d, e is divisible by 3, then their squares are congruent to 1 modulo 3, and hence the sum of their squares is congruent to 5 modulo 3. But there is no integer f with $f^2 \equiv 5 \pmod 3$. So for any 6-tuple in S, one of the first five must be divisible by 3. It follows that 3 divides every element of T.

If n is odd then $n^2 \equiv 1 \pmod 8$, and if n is even then $n^2 \equiv 0$ or $4 \pmod 8$. Hence, if all the five numbers a, b, c, d, e are odd, then $f^2 \equiv 5 \pmod 8$, which is not possible. So suppose that k of these five numbers are odd and the remaining are even. Then, $k \leq 4$ and $f^2 \equiv k \pmod 4$. This shows that $k = 0, 1$ or 4. If $k = 0$ or 1, then the product $abcdef$ is divisible by 16.

Let $k = 4$. Suppose that a, b, c, d are odd and e is even. Then, $a^2 + b^2 + c^2 + d^2 \equiv 4 \pmod 8$. If $e = 4e_1 + 2$, then $e^2 \equiv 4 \pmod 8$, and so $f^2 \equiv 4 + 4 \pmod 8$. Thus, f^2 is divisible by 8 and so f is divisible by 4 in this case. On the other hand, if $e = 4e_1$ then, again, f^2 is even and f is even. It follows that ef is divisible by 8, in either case.

We have proved that every element in T is divisible by 8 as well as 3, and hence by 24. Since $1^2 + 1^2 + 1^2 + 2^2 + 3^2 = 4^2$, we see that $(1, 1, 1, 2, 3, 4) \in S$. The product of these 6 numbers is 24. Thus, $24 \in T$, and is a common divisor of all members of T. Hence the required greatest common divisor is 24.

15. Let α and β be the roots of the quadratic equation $x^2 + mx - 1 = 0$, where m is an odd integer. Let $\lambda_n = \alpha^n + \beta^n$ for $n \geq 0$. Prove that λ_n is an integer and that $\gcd(\lambda_n, \lambda_{n+1}) = 1$.

Solution Since $\alpha^2 + m\alpha - 1 = 0$, we have $\alpha^{n+2} + m\alpha^{n+1} - \alpha^n = 0$ for $n \geq 0$. Adding this to a similar result for β, we get

$$\lambda_{n+2} + m\lambda_{n+1} - \lambda_n = 0, \quad \text{for } n \geq 0. \tag{*}$$

Now, $\lambda_0 = 1 + 1$ and $\lambda_1 = -m$ are integers. It follows from (*) by induction that λ_n is an integer for all n.

The principle used in the Euclidean algorithm for finding gcd is the following: Let a, b be integers and $b \neq 0$. If $a = bx + c$, where x and c are integers, then $(a, b) = (b, c)$. Using this, we see from (*) that $(\lambda_{n+2}, \lambda_{n+1}) = (\lambda_{n+1}, \lambda_n)$. Since $\lambda_0 = 2$, $\lambda_1 = -m$, and m is odd, we have $(\lambda_1, \lambda_0) = 1$. Hence, $(\lambda_{n+2}, \lambda_{n+1}) = 1$ by induction.

16. Let $p_1, p_2, \ldots$ be a sequence of primes defined by $p_1 = 2$, and, for $n \geq 1$, p_{n+1} is the largest prime divisor of $p_1 p_2 \cdots p_n + 1$. (Thus, $p_2 = 3, p_3 = 7$.) Prove that $p_n \neq 5$ for any n, and also that $p_n \neq 11$ for any n.

Solution Since $p_1 = 2$ and $p_2 = 3$, we see that for any $n \geq 2$, the number $N = p_1 p_2 \cdots p_n + 1$ is not divisible by 2 or 3. Hence, if $p_{n+1} = 5$, then the largest prime divisor of N is 5, and so $N = 5^k$. In that case, $p_1 p_2 \cdots p_n = 5^k - 1$ is to be divisible by 4. This is not possible.

Clearly, N is not divisible by 7 for $n \neq 3$ since $p_3 = 7$. Suppose that $p_{n+1} = 11$ for some n. Then, $n > 2$; and further, 5 and 11 are the only possible prime divisors of N, and hence $N = 5^k \cdot 11^r$. So $N \equiv 1^k(-1)^r \pmod 4$. If r is even, this implies that $N - 1$ is divisible by 4, which is not possible. Hence, r is odd. Further, $N \equiv 1 \pmod 3$, and

$$N = 5^k \cdot 11^r \equiv (-1)^k(-1)^r \pmod 3.$$

Thus, $1 \equiv (-1)^{k+r}$, and $k + r$ is even. So k is odd. Again,

$$N = 5^k \cdot 11^r \equiv (-2)^k(4)^r \pmod 7.$$

Since k is odd, this becomes $N \equiv (-1)2^{k+2r} \pmod 7$. Since $2^m \equiv 2$, 4 or 1 modulo 7, we get $N \equiv -2, -4$ or $-1 \pmod 7$. But $N \equiv 1 \pmod 7$ since $n > 2$. This gives a contradiction. So p_{n+1} cannot be 11.

17. Find all primes p and q such that $p(p+1)+q(q+1)=n(n+1)$ for some positive integer n.

Solution Let p, q, n be like this and $p \leq q$. Then, $n > q \geq p$, and

$$(p+q)(p+q+1) = p(p+1) + q(q+1) + 2pq = n(n+1) + 2pq,$$

which implies that $p + q > n$. Again,

$$q(q+1) = n^2 - p^2 + n - p = (n-p)(n+p+1). \qquad (1)$$

Hence, q divides $n - p$ or $n + p + 1$.

Let $q|(n-p)$ and $n - p = kq$. Then, $k \geq 1$ and $n = kq + p$. From (1),

$$q + 1 = k(n+p+1) = k(kq+2p+1) \geq q + 2p + 1 > q + 2.$$

This contradiction shows that $q|(n+p+1)$. Let $n+p+1 = kq$. Since $p + q > n > q \geq p$, we get

$$kq = n + p + 1 < p + q + p + 1 \leq 3q + 1,$$

and so $1 < k \leq 3$, that is, $k = 2$ or 3.

Further, $n = kq - p - 1$, and $q + 1 = k(n-p)$ from (1). Hence,

$$kp = kn - q - 1 = k^2q - kp - k - q - 1,$$
$$2kp = (k^2-1)q - k - 1. \qquad (2)$$

Let $k = 2$. Then (2) shows that $4p = 3q - 3$, and so $3|p$. Thus, $p = 3$, $q = 5$ and $n = 2q - p - 1 = 6$.

Next, let $k = 3$. Then, again from (2), $6p = 8q - 4$. This implies that $p = 2$, and hence, $q = 2$, $n = 3q - p - 1 = 3$.

We may also have $p \geq q$. Thus we get three solutions:

$$p = 3, q = 5, n = 6; \quad p = 5, q = 3, n = 6; \quad p = 2, q = 2, n = 3.$$

18. For a given positive integer n, let u_n and v_n be the number of binomial coefficients $\binom{n}{k}$ which are congruent modulo 3 to 1 and 2 respectively. Prove that $u_n > v_n$. Prove also that $u_n - v_n$ is a power of 2.

Solution For $n = 1, 2, 3$, we can make a direct count and see that

$$u_1 = 2,\ v_1 = 0; \quad u_2 = 2,\ v_2 = 1; \quad u_3 = 2,\ v_3 = 0.$$

First observe that if a polynomial congruence $f(x) \equiv 0 \pmod{p}$, where p is prime, is satisfied for all integer values of x, then the coefficients in $f(x)$ are congruent to 0 mod p.

Since $(1+x)^3 \equiv 1+x^3 \pmod{3}$, we have $(1+x)^{3n} \equiv (1+x^3)^n \pmod{3}$. This is satisfied by all integers x. So comparing the coefficients of x^k, we get

$$\binom{3n}{k} \equiv \begin{cases} \binom{n}{m} & \text{when } k = 3m \\ 0 & \text{when } k = 3m+1, 3m+2. \end{cases}$$

Let us see through the binomial coefficients $\binom{3n}{k}$, $0 \le k \le 3n$, and search for those which are congruent to $1, 2$ modulo 3. We get the same number as we go through the coefficients $\binom{n}{m}$, $0 \le m \le n$. It follows that

$$u_{3n} = u_n, \quad v_{3n} = v_n.$$

Similarly, since $(1+x)^{3n+1} \equiv (1+x^3)^n(1+x) \pmod{3}$, and $(1+x)^{3n+2} \equiv (1+x^3)^n(1+2x+x^2) \pmod{3}$, we get

$$\binom{3n+1}{k} \equiv \begin{cases} \binom{n}{m} & \text{when } k = 3m, 3m+1 \\ 0 & \text{when } k = 3m+2, \end{cases}$$

$$\binom{3n+2}{k} \equiv \begin{cases} \binom{n}{m} & \text{when } k = 3m, 3m+2 \\ 2\binom{n}{m} & \text{when } k = 3m+1. \end{cases}$$

These imply, as before, that

$$u_{3n+1} = 2u_n, \quad v_{3n+1} = 2v_n,$$

$$u_{3n+2} = 2u_n + v_n, \quad v_{3n+2} = 2v_n + u_n.$$

Thus, $u_{3n} - v_{3n} = u_n - v_n$, $u_{3n+1} - v_{3n+1} = 2(u_n - v_n)$ and $u_{3n+2} - v_{3n+2} = u_n - v_n$. It follows that if $u_n > v_n$, then $u_{3n} - v_{3n}$, $u_{3n+1} - v_{3n+1}$ and $u_{3n+2} - v_{3n+2}$ are positive. Since $u_1 - v_1 = 2$, $u_2 - v_2 = 1$ and $u_3 - v_3 = 2$ are positive, we conclude, by induction, that $u_n - v_n > 0$ for all n. The second statement also follows from this.

19. Let p be a prime. For a given positive integer n, let u_n be number of binomial coefficients $\displaystyle\binom{n}{k}$ which are not divisible by p. Prove that $u_{np+r} = (r+1)u_n$ for $0 \le r \le p-1$.

Solution Let $1 \le r \le p-1$. Since $(1+x)^p \equiv 1 + x^p \pmod{p}$, we have

$$\begin{aligned}(1+x)^{np+r} &\equiv (1+x^p)^n(1+x)^r \pmod{p}\\ &\equiv (1+x^p)^n\left\{1+\binom{r}{1}x+\binom{r}{2}x^2+\cdots\right\} \pmod{p}.\end{aligned}$$

This is a polynomial congruence satisfied by all integer values of x. So comparing the coefficients of x^k, where $k = mp + s$, $0 \le s \le r$, we see that

$$\binom{np+r}{k} \equiv \binom{n}{m}\binom{r}{s} \pmod{p}.$$

Consider the binomial coefficients $\binom{np+r}{k}$, $0 \le k \le np+r$, and search for those which are not congruent to 0 mod p. None of the $r+1$ numbers $\binom{r}{s}$, $0 \le s \le r$, is divisible by p. Hence we find the same number $\binom{n}{m}$ repeated $r+1$ times. It follows that $u_{np+r} = (r+1)u_n$. If $r = 0$, we have to consider the expansion of $(1+x)^{np}$. Since

$$(1+x)^{np} \equiv \sum_{m=0}^{n}\binom{n}{m}x^{mp} \pmod{p},$$

we get $\binom{np}{mp} \equiv \binom{n}{m} \pmod{p}$. This gives $u_{np} = u_n$, which completes the proof.

20. Let v_n be the number of even binomial coefficients $\binom{n}{r}$. Show that there exist an infinite number of values of n such that $v_n = 0$.

Solution Let u_n be the number of odd binomial coefficients $\binom{n}{r}$. Then, $v_n = n+1-u_n$. The previous problem with $p = 2$ shows that $u_{2n} = u_n$ and $u_{2n+1} = 2u_n$. Hence,

$$v_{2n+1} = 2n+2-u_{2n+1} = 2n+2-2u_n = 2v_n.$$

This shows that if $v_n = 0$, then $v_{2n+1} = 0$. Since $v_1 = 0$, we get $v_3 = 0$, $v_7 = 0$, and so on. In general, $v_m = 0$ whenever $m = 2^k - 1$.

21. Find the least possible value of $a+b$, where a, b are positive integers such that 11 divides $a + 13b$ and 13 divides $a + 11b$.

Solution By what has been given,

$$a + 24b = (a + 13b) + 11b = (a + 11b) + 13b$$

is divisible by 11 and 13, and hence by 143. So, $6a + b = 6(a + 24b) - 143b$ is divisible by 143. Let $6a+b = 143k$. If $k = 1$, then $(a+b)+5a = 143$. The solution of this with the least value for $a + b$ is $a = 23$, $b = 5$, giving $a + b = 28$. If $k \geq 2$, then $6(a + b) > 6a + b \geq 286$, and hence $a + b > 28$. So the least possible value of $a + b$ is 28.

22. Prove that there are infinitely many positive integers n such that $n(n + 1)$ can be expressed as a sum of two positive squares in *at least* two different ways. (Here $a^2 + b^2$ and $b^2 + a^2$ are considered as the same representation.)

    Solution Let x, y, z be a Pythagorean triple so that $x^2 + y^2 = z^2$. If $n = z^2$, then $n(n+1) = (x^2+y^2)(z^2+1)$. Thus, $n(n+1) = (z^2)^2+z^2$, and

    $$n(n + 1) = (xz + y)^2 + (yz - x)^2 = (yz + x)^2 + (xz - y)^2.$$

    Thus, every Pythagorean triple gives a value of n such that $n(n + 1)$ has at least two representations as a sum of two squares.

23. Prove that there exist infinitely many integer pairs (m, n) such that $m|(n^2 + 1)$ and $n|(m^2 + 1)$.

    Solution Suppose that (m, n) is a pair of positive integers such that $m|(n^2 + 1)$, $n|(m^2 + 1)$ and $m < n$. Clearly, $(1, 2)$ is such a pair. Let $k = (n^2 + 1)/m$. Then, k is a positive integer, $mk = n^2 + 1$, and hence $k|n^2 + 1$. Since $m^2 + 1 = an$ for some integer a, we have

    $$m^2(k^2 + 1) = (n^2 + 1)^2 + m^2 = n^4 + 2n^2 + an = n(n^3 + 2n + a).$$

    Since $mk = n^2 + 1$, we observe that n is relatively prime to m; hence, $n|(k^2 + 1)$. Further, $mk = n^2 + 1 > n^2 > mn$ and so $k > n$. Thus, we get a new pair (n, k) such that $n|(k^2 + 1)$, $k|(n^2 + 1)$ and $n < k$. Starting from $(1, 2)$, we can construct any number of pairs (n, k) such that $n|(k^2 + 1)$, $k|(n^2 + 1)$ and $n < k$. Thus, $(1, 2), (2, 5), (5, 13), \ldots$ satisfy the conditions.

24. If n is divisible by an odd prime and $n > 3$, prove that $2^n + 1$ has at least two distinct prime divisors.

Solution Let $n = mp$, where p is an odd prime. Then, $2^n+1 = 2^{mp}+1$ is composite as it is divisible by $2^m + 1$. Suppose that $2^n + 1$ is a prime power, and let $2^n+1 = q^k$, where q is an odd prime. Now, $q^{k-1}+\cdots+1$ is a divisor of $q^k - 1 = 2^n$, and this is the sum of k terms each of which is odd. Since the divisors of 2^n are $1, 2, 4, \ldots, 2^n$, we see that k is necessarily even. In that case, $2^n = (q^{k/2} - 1)(q^{k/2} + 1)$, and these two factors are even and differ by 2. So $q^{k/2} - 1 = 2$ and $q^{k/2} + 1 = 4$. Thus, $q^{k/2} = 3$, $q = 3$, $k = 2$, and $n = 3$. It follows that if $n > 3$, then $2^n + 1$ is a composite number which is not a prime power. So it must have two distinct prime divisors if $n > 3$.

25. Show that the number $\left(2^{2^r}\right)^n - 1$ has at least $2r + 1$ distinct prime divisors if $n > 2^r$, unless $r = 1, n = 3$.

Solution If $r = 0$, then $n > 1$, and $\left(2^{2^r}\right)^n - 1 = 2^n - 1 > 1$; the result is true in this case. So suppose that $r \geq 1$, $n > 2^r$, and $n > 3$. Let $a = 2^n$. Then,

$$\left(2^{2^r}\right)^n - 1 = a^{2^r} - 1 = (a-1)(a+1)\cdots(a^{2^{r-1}} + 1). \qquad (*)$$

The product of the first $k + 1$ factors in (*) is

$$(a-1)(a+1)\cdots(a^{2^{k-1}} + 1) = a^{2^k} - 1 \quad \text{for } 1 \leq k \leq r.$$

If b is even, then $\gcd(b - 1, b + 1) = 1$. So $a^{2^k} + 1$ is relatively prime to $a^{2^k} - 1$. It follows that each of the $r + 1$ factors in (*) is relatively prime to the product of the preceding factors. Consequently, the $r + 1$ factors in (*) have distinct prime divisors.

First suppose that n has an odd prime divisor. Now, $a^k + 1 = 2^{nk} + 1$, and nk has an odd prime divisor. Since $nk \geq n > 3$ for $k \geq 1$, we see, by the previous problem, that $a^k + 1$ has at least two distinct prime divisors for $k \geq 1$. It follows from (*) that $2^{n2^r} - 1$ has at least $1 + 2r$ distinct prime divisors.

Next suppose that n is a power of 2. Let $n = 2^k$. Then $k > r$, and

$$\left(2^{2^r}\right)^n = 2^{n2^r} = 2^{2^{r+k}} = F_{r+k} - 1,$$

where F_n is the nth Fermat number. We have $F_{n-1}(F_{n-1}-2) = F_n-2$, and hence, by induction, $F_n - 2 = F_0F_1 \cdots F_{n-1}$. Since the Fermat numbers $F_0, \ldots, F_{n-1}$ are pair-wise coprime, we see that $F_n - 2$ has at least n distinct prime divisors. Thus, $2^{n2^r} - 1 = F_{r+k} - 2$ has at least $r + k$ distinct prime divisors. Since $k > r$, we have $r + k \geq 2r + 1$.

26. Prove that (i) 3^{n+1} divides $2^{3^n}+1$ for all $n \geq 1$, and (ii) 3^{n+2} does not divide $2^{3^n}+1$ for any n.

Solution Both (i) and (ii) are true for $n=1$ since $2^{3^1}+1$ is divisible by 3^2 but not by 3^3. Assume that these hold for some $n \geq 1$. Let $a = 2^{3^n}$. Then, $a+1 = 2^{3^n}+1 = m3^{n+1}$ for some integer m, not divisible by 3. It follows that

$$\begin{aligned} 2^{3^{n+1}}+1 &= a^3+1 = (a+1)^3 - 3a(a+1) \\ &= m^3\,3^{3(n+1)} - 3\left(m\,3^{n+1}-1\right)m3^{n+1} = m_1\,3^{n+2}, \end{aligned}$$

where m_1 is an integer. Now,

$$m_1 = m^3 3^{2n+1} - m^2 3^{n+1} + m \equiv m \pmod 3,$$

Hence, m_1 is not divisible by 3 since m is not divisible by 3. Thus, (i) and (ii) hold for $n+1$. This proves the result by induction.

27. Let n be odd and $\gcd(a,n)=1$. Prove that if a^n+1 is divisible by n, then either $n=1$ or $\gcd(a+1,n)>1$.

Solution We have to prove that if n is an odd integer such that

$$n>1 \quad \text{and} \quad a^n+1 \equiv 0 \pmod n \Longrightarrow \gcd(a+1,n)>1. \tag{1}$$

If $n=3$, then we have

$$0 \equiv a^3+1 \equiv a+1 \pmod 3,$$

and hence, $\gcd(a+1,3)=3$. So the result is true in this case. Assume that it is true for all odd integers $m<n$, that is, if $1<m<n$, then

$$a^m+1 \equiv 0 \pmod m \Longrightarrow \gcd(a+1,m)>1. \tag{2}$$

Since n is odd, we have $(-a)^n = -a^n \equiv 1 \pmod n$. Clearly, a is relatively prime to n. So by Euler's theorem, $(-a)^{\phi(n)} \equiv 1 \pmod n$. Let m be the least positive integer such that $(-a)^m \equiv 1 \pmod n$. Then, $m|n$ and $m|\phi(n)$. Hence, m is odd, $m \leq \phi(n) < n$, and a^m+1 is divisible by n. Since a^m+1 is also divisible by m if it is divisible by n, we see that if $m>1$, then we have an odd integer m such that

$$1<m<n, \quad a^m+1 \equiv 0 \pmod m.$$

By (2), this shows that $\gcd(a+1,m)>1$ which, in turn, implies that $\gcd(a+1,n)>1$. If $m=1$, then $a+1$ is divisible by n, and $\gcd(a+1,n)=n>1$. Thus, (1) holds, in any case.

28. Determine all positive integers n for which n^2 is a divisor of $2^n + 1$.

Solution Suppose that $2^n + 1$ is divisible by n^2, and $n > 1$. So $2^n + 1$ is divisible by n. Clearly, n must be odd. The previous problem with $a = 2$ shows that $(3, n) > 1$, and hence $3|n$. Let $n = m3^r$, where $r \geq 1$ and m is not divisible by 3.

(i) We shall first prove that $r = 1$. Assume that $r \geq 2$. Now, $m3^r = n$ is odd, and so m is odd. Thus, m is of the form $6k \pm 1$, as it is not divisible by 3. Let

$$x = 2^{3^r}, \quad y = x^m \quad z = x^6.$$

Since $\phi(3^{r+2}) = 6(3^r)$, we get

$$z = 2^{6(3^r)} = 2^{\phi(3^{r+2})} \equiv 1 \pmod{3^{r+2}}. \tag{1}$$

Moreover,

$$y + 1 = 2^{m3^r} + 1 = 2^n + 1 \equiv 0 \pmod{n^2}.$$

Since $n^2 = (m^2 3^{r-2})3^{r+2}$, we get

$$y \equiv -1 \pmod{3^{r+2}}. \tag{2}$$

If $m = 6k + 1$, then

$$xz^k = x^{6k+1} = x^m = y \equiv -1 \pmod{3^{r+2}}$$

from (2), and so $-1 \equiv xz^k \equiv x \pmod{3^{r+2}}$ from (1). If $m = 6k - 1$, then

$$xy = x^{m+1} = x^{6k} = z^k \equiv 1 \pmod{3^{r+2}}$$

by (1), and so $1 \equiv xy \equiv -x \pmod{3^{r+2}}$ by (2). Thus, in either case, $x \equiv -1 \pmod{3^{r+2}}$. But we have seen in an earlier problem that $x + 1 = 2^{3^r} + 1$ cannot be divisible by 3^{r+2}, and so $x \not\equiv -1 \pmod{3^{r+2}}$. This contradiction shows that $r < 2$, that is, $r = 1$.

(ii) Thus, $n = 3m$, and $2^n + 1 = 8^m + 1$ is divisible by $n = 3m$ and hence by m. If $m > 1$, the previous problem shows that $(8+1, m) > 1$, which in turn implies that $3|m$. It follows that $m = 1$ and $n = 3$. Hence, $2^n + 1$ is divisible by n^2 if and only if $n = 1$ or 3.

29. Find all integers a, b, c with $1 < a < b < c$ such that $(a-1)(b-1)(c-1)$ is a divisor of $abc - 1$.

Solution Let $x = a-1$, $y = b-1$ and $z = c-1$. Then, $0 < x < y < z$. Suppose that $abc - 1$ is divisible by $(a-1)(b-1)(c-1)$. Then there is a positive integer N such that $abc - 1 = N(a-1)(b-1)(c-1)$, that is, such that

$$(1+x)(1+y)(1+z) - 1 = Nxyz,$$
$$x + y + z + xy + yz + zx = (N-1)xyz.$$

This first of all shows that

$$(N-1)xyz > yz, \quad (N-1)x > 1$$

and also that

$$\begin{aligned}(N-1)xyz &= (x+1)(y+1+z) + yz - 1\\ &\leq y(z+z) + yz - 1 < 3yz,\end{aligned}$$

and hence, $(N-1)x < 3$. Combining these two, we get $1 < (N-1)x < 3$, $(N-1)x = 2$, and so $x = 1$ or 2.

The case $x = 1$ gives: $1 + y + z + y + yz + z = (N-1)xyz = 2yz$, and hence, $(y-2)(z-2) = 5$. It follows that $y - 2 = 1$, $z - 2 = 5$, as $y < z$, and $y = 3$, $z = 7$.

Similarly, the case $x = 2$ gives: $2+y+z+2y+yz+2z = (N-1)xyz = 2yz$, and hence, $(y-3)(z-3) = 11$. Thus, $y - 3 = 1$, $z - 3 = 11$, $y = 4$, $z = 14$.

We get two solutions: (i) $a = 2, b = 4, c = 8$, (ii) $a = 3, b = 5, c = 15$.

30. The increasing sequence $1, 3, 4, 9, 10, 12, 13, \ldots$ consists of positive integers which are powers of 3 or sums of distinct powers of 3. What is the 1000th term in it?

Solution Let x_n be the nth term. Then,

$$x_n = a_0 + a_1 \cdot 3 + a_2 \cdot 3^2 + \cdots + a_k \cdot 3^k,$$

where $a_i = 0$ or 1 for each i, and $a_k = 1$. Every integer n has a unique binary representation

$$n = a_0 + a_1 \cdot 2 + a_2 \cdot 2^2 + \cdots + a_k \cdot 2^k,$$

where $a_i = 0$ or 1 for each i, and $a_k = 1$. Let

$$f(n) = \sum_{i=0}^{k} a_i\, 3^i, \quad \text{when } n = \sum_{i=0}^{k} a_i\, 2^i,\ a_i = 0 \text{ or } 1, a_k = 1.$$

Thus,

$$f(1) = 1 = x_1,\ f(2) = 3^1 = x_2,\ f(3) = 1 + 3^1 = x_3, \ldots.$$

Suppose that $m = \sum_{i=0}^{r} b_i\, 2^i$, $b_i = 0$ or 1, $b_r = 1$, and $m < n$. Then,

$$2^r \le m < n \le 1 + 2 + 2^2 + \cdots + 2^k < 2^{k+1},$$

which shows that $r < k + 1$. If $r < k$, then

$$f(m) \le 1 + 3 + 3^2 + \cdots + 3^r < 3^{r+1} \le 3^k \le f(n).$$

Let $r = k$. We have $a_k = 1 = b_k$. If $a_i = b_i$ for all i, then $m = n$. Suppose that $a_i = b_i$ for $j < i \le k$ and $a_j \ne b_j$. Since $|a_i - b_i| \le 1$ for all i, we get

$$n - m = \sum_{i=0}^{j} (a_i - b_i) 2^i \le \sum_{i=0}^{j-1} 2^i + (a_j - b_j) 2^j = 2^j - 1 + (a_j - b_j) 2^j.$$

Since $a_j - b_j = \pm 1$, and $n > m$, this shows that $a_j - b_j = 1$. Thus,

$$\begin{aligned} f(n) - f(m) = \sum_{i=0}^{j} (a_i - b_i) 3^i &\ge \sum_{i=0}^{j-1} (-1) 3^i + (a_j - b_j) 3^j \\ &= -\frac{3^j - 1}{2} + 3^j > 0. \end{aligned}$$

It follows that $f(m) < f(n)$, whenever $m < n$. So $\{f(n)\}$ and $\{x_n\}$ are strictly increasing sequences, and $f(n) = x_n$ when $n = 1, 2, 3$. Since each term in the sequence is $f(n)$ for some n, we see that $f(n) = x_n$ for all n. So the 1000th term is $f(1000)$. Since $1000 = 2^3 + 2^5 + 2^6 + 2^7 + 2^8 + 2^9$, we have $f(1000) = 3^3 + 3^5 + 3^6 + 3^7 + 3^8 + 3^9 = 29\,430$.

where $0 \leq a_i \leq 4$, $a_k \neq 0$. Then,

$$[n/5] = a_1 + a_2 \cdot 5 + a_3 \cdot 5^2 + \cdots + a_k \cdot 5^{k-1},$$
$$\left[n/5^2\right] = a_2 + a_3 \cdot 5 + \cdots + a_k \cdot 5^{k-2},$$

and so on. Adding,

$$e = a_1 + a_2(5+1) + \cdots + a_k\left(5^{k-1} + 5^{k-2} + \cdots + 1\right).$$

If $1 + 5 + \cdots + 5^i = b_i$, then

$$b_1 = 6, \quad b_2 = 31, \quad b_3 = 156, \quad b_4 = 781, \ldots.$$

We have to find n such that $e = 1993$, that is,

$$1993 = a_1 + 6a_2 + 31a_3 + 156a_4 + 781a_5 + \cdots.$$

Since $b_5 > 5^5 > 1993$ and $2 \times 781 < 1993 < 3 \times 781$, we see that $k = 5$, $a_5 = 2$, and

$$a_1 + 6a_2 + 31a_3 + 156a_4 = 1993 - 2 \times 781 = 431.$$

We find successively, that $a_4 = 2$, $a_3 = 3$, $a_2 = 4$ and $a_1 = 2$. Hence, we have

$$n = a_0 + 2 \cdot 5 + 4 \cdot 5^2 + 3 \cdot 5^3 + 2 \cdot 5^4 + 2 \cdot 5^5 = a_0 + 7985,$$

where $0 \leq a_0 \leq 4$. If n has any of these five values, then $n!$ ends in exactly 1993 zeros.

5. Find the highest power of 1991 that divides the number

$$1990^{1991^{1992}} + 1992^{1991^{1990}}.$$

Solution Let $a = 1991$, and

$$N_1 = (a-1)^{a^{a+1}}, \quad N_2 = (a+1)^{a^{a-1}}.$$

We have to find the highest power of a that $N_1 + N_2$. We shall first prove that if $n = a^b$ and $k \geq 2$, then $\binom{n}{k} a^k$ is divisible by a^{b+2}.

We have $a = 11 \times 181$. For $k \geq 2$, let $k! = 11^r \cdot 181^s \cdot u$, where u is an integer not divisible by 11 or 181. Then,

$$r = \left[\frac{k}{11}\right] + \left[\frac{k}{11^2}\right] + \cdots.$$

We have a similar expression for s which shows that $s < r$. Thus,

$$\binom{n}{k} = \frac{n(n-1)\cdots(n-k+1)}{k!} = \frac{n(n-1)\cdots(n-k+1)}{11^r\, 181^s\, u}.$$

The value of this quotient is an integer. Since u is relatively prime to a and hence to $a^b = n$, we see that u divides the product $(n-1)\cdots(n-k+1)$. If this product is equal to uu_1, then

$$\binom{n}{k} a^k = \frac{nuu_1}{11^r\, 181^s\, u} 11^k\, 181^k = nu_1 a^{k-r} 181^{r-s},$$

since $a = 11 \cdot 181$. This is divisible by $na^{k-r} = a^{b+k-r}$. Now,

$$r \leq \frac{k}{11} + \frac{k}{11^2} + \cdots = \frac{k}{10}.$$

If $k = 2$, then $r = 0$, and $k - r = 2$. If $k \geq 3$, then $k - r \geq k - k/10 = 9k/10 > 2$. Thus, we have $k - r \geq 2$ for all $k \geq 2$. This proves our claim.

If we take $b = a + 1$, then $n = a^{a+1}$. Since n is odd, we get

$$N_1 = (-1 + a)^n = -1 + na + m_1\, a^{a+3} = -1 + a^{a+2} + m_1\, a^{a+3},$$

for some integer m_1. If $b = a - 1$, then $n = a^{a-1}$. We get

$$N_2 = (1 + a)^n = 1 + na + m_2 a^{a+1} = 1 + a^a + m_2 a^{a+1},$$

for some integer m_2. Combining these two, we see that $N_1 + N_2 = a^a + m_3 a^{a+1}$ for some integer m_3. This shows that the highest power $a = 1991$ that divides $N_1 + N_2$ is $a^a = 1991^{1991}$.

6. Find all natural numbers n such that the product of their digits (in decimal notation) is equal to $n^2 - 10n - 22$.

Solution Let $P(n)$ denote the product of digits in n. We have to find n such that $P(n) = n^2 - 10n - 22$. If n is a one-digit number, then $P(n) = n$, and this becomes $n = n^2 - 10n - 22$. There is no integer n

satisfying this equation. So suppose that n has k digits, $k \geq 2$, and let $n = a_0\,10^{k-1} + a_1\,10^{k-2} + \cdots + a_{k-1}$, where $a_0 \geq 1$. Then,

$$P(n) = a_0(a_1 \cdots a_{k-1}) \leq a_0 9^{k-1} < a_0 10^{k-1} \leq n.$$

So we must have $n^2 - 10n - 22 < n$, that is, n lies between the roots of $n^2 - 11n - 22 = 0$. Hence, $n < 14$, that is, $n = 10, 11, 12$ or 13. Of these, only $n = 12$ satisfies the condition.

7. Let n be an integer greater than 2, and V_n be the set of integers $1 + kn$, $k = 1, 2, \ldots$. A number $m \in V_n$ is called *indecomposable* in V_n if it is not the product of two elements in V_n. Prove that there exists an element $N \in V_n$ that can be expressed as the product of elements indecomposable in V_n in more than one way.

Solution If an element in V_n is not indecomposable in V_n, then it is the product of two elements in it. Repeated application of this fact to each of the factors, shows by induction that every element in V_n is either indecomposable or a product of indecomposables. We have to prove that such a factorisation is not unique.

Let $x = n - 1$, $y = 2n - 1$, and $N = x^2y^2$. Then, x^2, y^2 and xy are elements in V_n. If $x^2 = ab$, then $a \leq x$ or $b \leq x$. But the least element of V_n is $n + 1 > x$. So x^2 cannot be expressed as the product of two elements in V_n, that is, x^2 is indecomposable in V_n. Hence, N can be expressed as a product of indecomposable factors, one of which is x^2. Note that $N = (xy)(xy)$, and xy can be expressed as a product of indecomposable factors $z_1, \ldots, z_k$ in V_n. If one of these factors, say z_1, is x^2, then

$$\frac{xy}{x^2} = z_2 \cdots z_k$$

is an integer. But $xy/x^2 = y/x = 2 + 1/(n-1)$ is not an integer since $n > 2$. So the factorisation of xy into indecomposables gives another factorisation N without using x^2. This proves that the factorisation is not unique.

8. Let m and n be distinct natural numbers. Suppose that, in their decimal representations, the last three digits of 1978^m are equal, respectively to the last three digits of 1978^n. Find m, n such that $m + n$ has the least value.

Solution Let $m < n$. If the last three digits of 1978^m and 1978^n are equal, then their difference $1978^n - 1978^m$ is a multiple of 1000, that is, $1978^m(1978^{n-m} - 1)$ is divisible by 8 and 125. This implies that $m \geq 3$, and that $(1978^{n-m} - 1)$ is divisible by 125. Since

$$m + n = (n - m) + 2m \geq n - m + 6,$$

we shall choose m, n such that $n - m$ is as small as possible.

Let $a = 1978$, and h be the least positive integer such that $a^h - 1$ is divisible by 125. Since $\phi(125) = 125(1 - 1/5) = 100$, we have $a^{100} \equiv 1 \pmod{125}$ the by Euler–Fermat theorem. Further, if $a^h - 1$ is divisible by 125, then it is also divisible by 5. Since $a \equiv 3 \pmod 5$, we must have $3^h \equiv 1 \pmod 5$. Considering the powers $3, 3^2, 3^3, 3^4$, we see that 3^4 is the smallest power of 3 that is $\equiv 1 \pmod 5$. Combining these two facts, we see that h is a divisor of 100 and a multiple of 4. Consequently, the possible values of h are 4, 20 and 100.

Observe that $a^h - 1$ is also divisible by 25. Now, $a \equiv 3 \pmod{25}$, and hence,

$$a^4 \equiv 3^4 \equiv 6 \pmod{25}.$$

This first of all shows that $h \neq 4$. It also shows that $a^4 = 6 + 25k$ for some integer k. Hence,

$$\begin{aligned} a^{20} = (1 + 5 + 25k)^5 &= 1 + 5(5 + 25k) + 10(5 + 25k)^2 + \cdots \\ &\equiv 1 + 25 \pmod{125}. \end{aligned}$$

This rules out the possibility of $h = 20$. Thus, $h = 100$ which implies that $n - m \geq 100$. Thus, the least value of $m + n = (n - m) + 2m$ is 106.

9. Show that there exist infinitely many pairs (a, b) of positive integers such that

   (i) $ab(a + b)$ is not divisible by 7,

   (ii) $(a + b)^7 - a^7 - b^7$ is divisible by 7^7.

Solution It can be seen that

$$(a + b)^7 - a^7 - b^7 = 7ab(a + b)(a^2 + ab + b^2)^2.$$

Since $ab(a + b)$ should not be divisible by 7, we have to find a, b such that $a^2 + ab + b^2$ is divisible by $7^3 = 343$. If we get one such pair a, b, then we can get any number of pairs from it by adding multiples of 7^3 to a and b. So it is enough if we get one such pair.

Since $(a-b)(a^2+ab+b^2) = a^3-b^3$, we see that if $a-b$ is not divisible by 7 and $a^3 - b^3$ is divisible by 7^3, then $a^2 + ab + b^2$ is divisible by 7^3.

To find one such pair a, b like this, choose $b = 1$. If $a \equiv 2$ or 4 modulo 7, then $a^3 \equiv b^3 \pmod 7$. Try with $a \equiv 4 \pmod 7$. Then, $a = 7n+4$,

$$4^3 + 3(4^2)(7n) + 3(4)(7n)^2 \equiv 1 \pmod{7^3},$$

and hence, $9 + 48n + 84n^2 \equiv 0 \pmod{49}$. This, first of all, implies that $48n + 9 \equiv 0 \pmod 7$. Thus,

$$n \equiv 2 \pmod 7, \quad a = 7(7m+2) + 4 = 49m + 18.$$

So m must satisfy $18^3 + 3(18)^2(49m) \equiv 1 \pmod{7^3}$. This is satisfied by $m = 0$. Thus we get one solution, namely, $a = 18, b = 1$. Moreover, $a = 7^3k + 18, b = 7^3r + 1$ are solutions for all positive integers k, r.

10. Let a, b, c and d be odd integers such that $0 < a < b < c < d$ and $ad = bc$. Prove that if $a + d = 2^k$ and $b + c = 2^m$ for some positive integers k and m, then $a = 1$. Determine all such integers a, b, c, d.

Solution Since $ad = bc$, we have $(a-b)(a-c) = a(a-b-c+d) = a(2^k - 2^m)$. The l.h.s. is positive, so we get $k > m$. Since $b - a$ and $c - a$ are even positive integers, we have

$$b - a = 2^r x, \quad c - a = 2^s y,$$

where $r \geq 1$, $s \geq 1$, and x, y are odd. Then,

$$(2^r x)(2^s y) = (2^k - 2^m)a = 2^m(2^{k-m} - 1)a.$$

This implies that $r + s = m$. Further,

$$2^r x + 2^s y = (b-a) + (c-a) = 2^m - 2a.$$

$$2^{r-1}x + 2^{s-1}y + a = 2^{m-1}. \qquad (*)$$

Since $2^m = b + c \geq 3 + 5$, we have $m \geq 3$, and hence 2^{m-1} is even. As $x \equiv y \equiv a \equiv 1 \pmod 2$, we see from (*) that $2^{r-1} + 2^{s-1} + 1 \equiv 0 \pmod 2$. Hence, we must have $r = 1$ or $s = 1$. If $s = 1$, then $r = m-1$ and so

$$b > 2^r x \geq 2^{m-1} = \frac{b+c}{2},$$

which implies that $b > c$. It follows that $r = 1$ and $s = m - 1$. So (*) gives

$$x + a = 2^{m-1} - 2^{m-2}y = 2^{m-2}(2 - y).$$

This shows that $y < 2$, that is, $y = 1$. Hence

$$c = a + 2^s = 2^{m-1} + a, \quad b = 2^m - c = 2^{m-1} - a.$$
$$ad = bc = 2^{2m-2} - a^2.$$

This shows that 2^{2m-2} is divisible by a. Since a is odd, this forces $a = 1$. Thus, if a, b, c, d satisfy the given conditions, then

$$a = 1, \quad b = 2^{m-1} - 1, \quad c = 2^{m-1} + 1, \quad d = bc = 2^{2m-2} - 1,$$

where $m \geq 3$. Clearly, if a, b, c, d are like this, then they satisfy the given conditions with $k = 2m - 2$.

11. Show that there exist infinitely many sets of 1983 consecutive positive integers each of which is divisible by some number of the form a^{1983} with $a \geq 2$.

Solution Write $m = 1983$. We shall prove, by induction on n, that there exist a set of n consecutive integers each of which is divisible by a number of the form a^m with $a > 1$.

If $n = 1$, choose any $a > 1$ and consider the one-element set $\{a^m\}$.

Assume that there exist n consecutive integers $N, N + 1, \ldots$ such that $N + i - 1$ is divisible by a_i^m, $a_i > 1$, $1 \leq i \leq n$. Let

$$P = (a_1 a_2 \cdots a_n)^m, \quad Q = (P + 1)^m - 1 \quad N_1 = (N + n)Q.$$

It is clear that Q is divisible by P, and hence by each a_i^m. So

$$N_1 + (N + i - 1) = (N + n)Q + (N + i - 1)$$

is divisible by a_i^m for $1 \leq i \leq n$. Finally,

$$N_1 + N + n = (N + n)(Q + 1)$$

is divisible by $Q + 1 = (P + 1)^m$. Thus, we get $n + 1$ integers

$$N_1 + N, N_1 + N + 1, \ldots, N_1 + N + n,$$

each of which is divisible by some mth power.

12. Each of the numbers $x_1, x_2, \ldots, x_n$ equals 1 or -1. If

$$\sum_{i=1}^{n} x_i x_{i+1} x_{i+2} x_{i+3} = 0,$$

where $x_{n+1} = x_1$, $x_{n+2} = x_2$ and $x_{n+3} = x_3$, prove that n is divisible by 4.

Solution Let $a_k = x_k x_{k+1} x_{k+2} x_{k+3}$ for $k = 1, 2, \ldots, n$. By hypothesis, $a_k = \pm 1$ for each k, and $a_1 + a_2 + \cdots + a_n = 0$. Suppose that $a_k = -1$ for m values of k and $a_k = 1$ for the remaining $n - m$ values. Then, the sum is equal to $-m + (n - m) = n - 2m$. Hence, $n = 2m$. The product of $a_1, a_2, \ldots, a_n$ is equal to $(-1)^m$. In this product, each x_i appears exactly four times. So the value of the product must be 1. Hence, $(-1)^m = 1$. Thus, m is even, and $n = 2m$ is divisible by 4.

13. Factor the number $5^{1985} - 1$ into a product of three integers each of which is larger than 5^{100}.

Solution We have $1985 = 5 \times 397$, and

$$x^5 - 1 = (x - 1)(x^4 + x^3 + x^2 + x + 1).$$

Take $x = 5^{397}$. We get one factor, namely, $5^{397} - 1$ which is greater than 5^{100}. To get other factors, note that

$$x^4 + x^3 + x^2 + x + 1 = (x^2 + 3x + 1)^2 - 5x(x + 1)^2.$$

If $x = 5^{397}$, then $5x(x + 1)^2 = a^2$, where $a = 5^{199}(5^{397} + 1)$. Writing $b = x^2 + 3x + 1$, we get $5^{1985} - 1 = (x - 1)(b + a)(b - a)$. Using the fact that $\sqrt{x} > \sqrt{5}$, we get

$$x^2 > x\sqrt{5x}, \quad 3x > 3\sqrt{5x},$$
$$b > \sqrt{5x}(x + 3) = a + 2\sqrt{5x} > a + 5^{100}.$$

Thus, $b - a > 5^{100}$, and so each of the three factors is larger than 5^{100}.

14. If $(1 + x + x^2 + x^3 + x^4)^{496} = a_0 + a_1 x + \cdots + a_{1984} x^{1984}$,

(a) determine the gcd of the coefficients $a_3, a_8, a_{13}, \ldots, a_{1983}$;

(b) show that $10^{347} > a_{992} > 10^{340}$.

Solution (a) Let ω be a non-real fifth root of unity. Then, $1, \omega, \omega^2, \omega^3$ and ω^4 are the distinct roots of the equation $x^5 - 1 = 0$. The sum of these roots is zero. We have

$$x^2(1 + x + x^2 + x^3 + x^4)^{496} = x^2\left(a_0 + a_1 x + \cdots + a_{1984} x^{1984}\right).$$

Note that $a_3, a_8, \ldots$ are now the coefficients of $x^5, x^{10}, \ldots$. Replace x by each of the five roots, and add the resulting equations. Using the fact that $\omega^5 = 1$ and $1 + \omega + \omega^2 + \omega^3 + \omega^4 = 0$, we see that

$$5^{496} = 5(a_3 + a_8 + a_{13} + \cdots).$$

The gcd of $a_3, a_8, \ldots$ divides their sum. So the gcd must be a power of 5. It can be seen that $a_{1983} = 496$, and hence the gcd is not divisible by 5. Thus, the gcd is 1.

(b) The expansion of $(1 + x + \cdots + x^4)^2$ is

$$1 + 2x + 3x^2 + 4x^3 + 5x^4 + 4x^5 + 3x^6 + 2x^7 + x^8.$$

The coefficients are symmetric about the middle. They increase in the beginning, attain the greatest value at the centre and then decrease. We shall prove that the same pattern holds for the coefficients in the expansion

$$(1 + x + \cdots + x^4)^n = b_0 + b_1 x + \cdots + b_{4n} x^{4n}.$$

Change x to $1/x$ and multiply by x^{4n}. We get,

$$(x^4 + x^3 + \cdots + 1)^n = b_0 x^{4n} + b_1 x^{4n-1} + \cdots b_{4n}.$$

Comparing these two, we find the symmetry of the coefficients, that is, $b_i = b_{4n-i}$ for all i.

Suppose that $b_i \le b_j$ if $i < j \le 2n$. If $\sum_{r=0}^{4n+4} c_r x^r$ is the sum when n is changed to $n+1$, then

$$\sum_{r=0}^{4n+4} c_r x^r = (1 + x + \cdots + x^4) \sum_{r=0}^{4n} b_r x^r.$$

It follows that

$$c_k = b_k + b_{k-1} + b_{k-2} + b_{k-3} + b_{k-4}, \qquad (*)$$

where we take $b_i = 0$ if $i < 0$ or $i > 4n$. This shows that $c_k - c_{k-1} = b_k - b_{k-5}$. Hence, $c_k - c_{k-1} \ge 0$ for $k \le 2n$, and

$$c_{2n+1} - c_{2n} = b_{2n+1} - b_{2n-4} = b_{2n-1} - b_{2n-4} \ge 0,$$

$$c_{2n+2} - c_{2n+1} = b_{2n+2} - b_{2n-3} = b_{2n-2} - b_{2n-3} \ge 0.$$

Thus we see that $c_i \leq c_j$ for $i < j \leq 2n + 2$. It follows, by induction on n, that the pattern holds for all n.

Next observe that the sum of all the coefficients is less than $(4n + 1)$ times the greatest one. Taking $x = 1$, we see that the sum is 5^n. From (*), we also see that if $b_k \leq 5^{n-1}$ for all k, then $c_k < 5^n$. The greatest coefficient when $n = 2$ is 5. It follows by induction that the greatest coefficient is less than 5^{n-1}.

Here, $n = 496$. The greatest coefficient is a_{992}. Thus, $5^{496} < 1985\, a_{992}$ and $a_{992} < 5^{495}$. Since $5^4 > 2^9$, we get

$$5^{496} = 5^{340} \cdot 5^{156} > 5^{340} \cdot 2^{351} = 10^{340} \cdot 2^{11}.$$

Thus, $1985\, a_{992} > 5^{496} > 1985 \cdot 10^{340}$, and hence, $a_{992} > 10^{340}$. Since $5^3 < 2^7$, we get

$$5^{495} = 5^{347} \cdot 5^{148} < 5^{347} \cdot 2^{148(7/3)} < 5^{347} \cdot 2^{347}.$$

Hence, $a_{992} < 10^{347}$.

15. Let k be a positive integer, $a_0 = 0$ and, for $n = 0, 1, 2, \ldots$,

$$a_{n+1} = k(a_n + 1) + (k + 1)a_n + 2\sqrt{k(k + 1)a_n(a_n + 1)}.$$

Prove that a_n is a positive integer for $n = 1, 2, 3, \ldots$.

Solution The given relation with $n = 0$ gives $a_1 = k$, and hence, $n = 1$ gives

$$a_2 = k(k + 1) + (k + 1)k + 2\sqrt{k(k + 1)k(k + 1)} = 4k(k + 1).$$

Moreover,

$$a_{n+1} = \left(\sqrt{(k + 1)a_n} + \sqrt{k(a_n + 1)}\right)^2,$$

$$a_{n+1} + 1 = \left(\sqrt{ka_n} + \sqrt{(k + 1)(a_n + 1)}\right)^2.$$

This implies that

$$\sqrt{k + 1}\sqrt{a_{n+1}} - \sqrt{k}\sqrt{a_{n+1} + 1} = (k + 1 - k)\sqrt{a_n}.$$

Squaring,

$$(k + 1)a_{n+1} + k(a_{n+1} + 1) - 2\sqrt{k(k + 1)a_{n+1}(a_{n+1} + 1)} = a_n.$$

Change n to $n+1$ in the given relation. Combining the result with this, we get

$$a_{n+2} + a_n = 2(k+1)a_{n+1} + 2k(a_{n+1}+1).$$

Since $a_1 = k$ and $a_2 = 4k(k+1)$ are positive integers, this proves by induction that a_n is an integer for all n. From the given relation,

$$a_{n+1} \geq k(a_n+1) + (k+1)a_n > a_n.$$

It follows that $a_n > 0$ for all $n \geq 1$.

16. Let d be any positive integer not equal to 2, 5 or 13. Show that one can find distinct a, b in the set $\{2, 5, 13, d\}$ such that $ab - 1$ is not a square.

Solution Suppose that $2d - 1 = x^2$, $5d - 1 = y^2$ and $13d - 1 = z^2$, where x, y, z are positive integers. Since $x^2 = 2d - 1$ is odd, it must be of the form $8k+1$, and hence, d is odd. So $y^2 = 5d-1$ and $z^2 = 13d-1$ are even. Let $y = 2y_1$ and $z = 2z_1$. Then, $8d = z^2 - y^2 = 4(z_1^2 - y_1^2)$. Hence, $(z_1 - y_1)(z_1 + y_1) = 2d$. But $(z_1 - y_1)$ and $(z_1 + y_1)$ are either both even or both odd. So their product cannot be equal to $2d$, where d is odd. This contradiction shows that at least one of $2d - 1$, $5d - 1$, $13d - 1$ is not a perfect square. Hence, we can take $a = 2, 5$ or 13, and $b = d$.

17. Let p be a prime. Prove that if $f(n) = n^2 + n + p$ is prime for all integers n such that $0 \leq n \leq \sqrt{p/3}$, then $f(n)$ is prime for all integers n such that $0 \leq n \leq p - 2$.

Solution Assume that $f(n)$ is prime for $0 \leq n < \sqrt{p/3}$, and also that $f(m)$ is composite for some non-negative $m \leq p - 2$. Choose the least such m. Now, $f(m)$ has a prime divisor $q \leq \sqrt{f(m)}$. Since $m \geq \sqrt{p/3}$, by assumption, we get

$$q^2 \leq f(m) = m^2 + m + 3\,(p/3) \leq m^2 + m + 3m^2 < (2m+1)^2.$$

Hence, $q < 2m + 1$. Further,

$$\begin{aligned} f(m) - f(n) &= (m^2 + m + p) - (n^2 + n + p) \\ &= (m - n)(m + n + 1). \end{aligned}$$

If $q \leq m$, let $n = m - q$ so that $m - n = q$, and $n < m$. If $q \geq m+1$, let $n = q - (m+1)$ so that $m + n + 1 = q$, and $n < (2m+1) - (m+1) = m$. In either case, we get a value of n such that $0 \leq n < m$, and

$f(m) - f(n)$ is divisible by q. It follows that $f(n)$ is divisible by q, and hence composite. This contradicts the choice of m, and proves the result.

It can be verified that the condition of the problem is satisfied when $p = 5, 11, 17$ or 41. If $p = 41$, the polynomial gives primes for 40 consecutive values of n.

18. (a) Let a and b be positive integers such that $ab + 1$ divides $a^2 + b^2$. Show that $\dfrac{a^2+b^2}{ab+1}$ is the square of an integer.

(b) Let k be an integer > 1. Show that we can find any number of pairs (a, b) of positive integers such that $\dfrac{a^2+b^2}{ab+1} = k^2$. Also show that $a \geq k$ and $b \leq a^3$ for all such pairs.

Solution (a) We shall prove a slightly more general result: If a, b and m are positive integers such that

$$0 \leq a^2 + b^2 - mab \leq m, \qquad (*)$$

then $a^2 + b^2 - mab$ is the square of an integer. This will show that if $\dfrac{a^2+b^2}{ab+1} = m$, then $a^2 + b^2 - mab = m$ is the square of an integer.

First suppose that $a = b$, then the first inequality in (*) shows that $2a^2 - ma^2 \geq 0$. Hence, $m = 1$ or 2, and $a^2 + b^2 - mab = a^2$ or 0. In either case, it is a square. So we need consider only the case $a \neq b$.

We proceed by induction on $n = \max(a, b)$. If $\max(a, b) = 1$, then $a = b = 1$, and so the result is true.

Assume that the result holds for all pairs (a, b) of positive integers for which $\max(a, b) < n$. Suppose that we have a pair (a, b) such that $\max(a, b) = n$, and (*) holds for some positive integer m. Because of the symmetry in a, b, we may suppose that $a < b$. Then $mab \leq a^2 + b^2 < 2b^2$, and hence $ma < 2b$. Let $c = ma - b$. Then $c < b$, $b + c = ma$, and so

$$b^2 - c^2 = ma(b - c), \quad b^2 - mab = c^2 - mac,$$

$$a^2 + b^2 - mab = a^2 + c^2 - mac.$$

This together with the second inequality in (*) implies that

$$m \geq a^2 + c^2 - mac, \quad m + mac \geq a^2 + c^2 > 0.$$

It follows that $c > -1$. If $c = 0$, then $a^2 + b^2 - mab = a^2$, a perfect square. Let $c > 0$. Since $a < b$ and $c < b$, we get a pair (a, c) of positive integers with $\max(a, c) < b = n$. Hence, $a^2 + c^2 - mac$ is a perfect square by induction hypothesis. Thus, $a^2 + b^2 - mab = a^2 + c^2 - mac$ is a perfect square. This completes the proof.

(b) Suppose that a, b, c are integers such that $a + c = k^2 b$. Then, $a^2 - c^2 = k^2 b(a - c)$, and as in (a),

$$a^2 + b^2 - k^2(ab + 1) = b^2 + c^2 - k^2(bc + 1),$$

$$\frac{a^2 + b^2}{ab + 1} = k^2 \Longrightarrow \frac{b^2 + c^2}{bc + 1} = k^2.$$

Further, $c = k^2 b - a \geq 4b - a$, and hence,

$$b > a \Longrightarrow c > b.$$

Define a_i for $i \geq 0$ as follows:

$$a_0 = k, \quad a_1 = k^3, \quad a_{i+1} = k^2 a_i - a_{i-1} \quad \text{for } i \geq 1.$$

Then, $a_0 < a_1$ and

$$a_0^2 + a_1^2 = k^2(a_0 a_1 + 1).$$

From what has been observed above, we see that $a_i < a_{i+1}$, and

$$a_i^2 + a_{i+1}^2 = k^2(a_i a_{i+1} + 1),$$

for all $i \geq 0$. Thus, (a_i, a_{i+1}) are pairs satisfying the condition. The sequence $a_0, a_1, a_2, \ldots$ corresponding to $k = 2$ is

$$2, 8, 30, 112, 418, \ldots.$$

If a, b are consecutive terms of this sequence, then $a^2 + b^2 = 4(ab + 1)$. Finally, suppose that $a^2 + b^2 = k^2(ab + 1)$. Then,

$$k^2 a(b + 1) = k^2 ab + k^2 a \geq k^2 ab + k^2 = a^2 + b^2 > b^2 - 1,$$

and hence, $k^2 a > b - 1$. This implies that $k^2 a \geq b$. Thus,

$$a^2 - k^2 = b(k^2 a - b) \geq 0,$$

and so

$$a \geq k, \quad a^3 \geq k^2 a \geq b.$$

This completes the proof.

19. Let $n > 6$ be an integer and $a_1, a_2, \ldots, a_k$ be all the natural numbers less than n and relatively prime to n. If

$$a_2 - a_1 = a_3 - a_2 = \cdots = a_k - a_{k-1} > 0,$$

prove that n must be either a prime or a power of 2.

Solution We have $a_1 = 1$ and $a_k = n - 1$. Let $d = a_2 - a_1$ and $S = \{a_1, a_2, \ldots, a_k\}$. If $d = 1$, then S contains all positive integers less than n, and hence n is prime. If $d = 2$, then $S = \{1, 3, \ldots, n-1\}$, that is, every odd integer less than n is relatively prime to n. So n cannot have an odd prime divisor, that is, n is a power of, 2 if $d = 2$.

Next suppose that $d > 2$. Then $a_2 = 1 + d > 3$, and hence 2 and 3 are not relatively prime to n. This implies that $2|n$ and $3|n$. If $5|n$, then $4|\phi(n)$ and so $\phi(n) \geq 4$. If 5 does not divide n and $n > 6$, then 1, 5 and $n - 1$ are less than n and relatively prime to n; in that case, $\phi(n) \geq 3$. This shows that if $n > 6$, then $k = \phi(n) > 2$. So S contains more than two elements; each of them is relatively prime to n. If $d \equiv 1 \pmod 3$, then $1 + 2d \equiv 0 \pmod 3$, that is, a_3 is divisible by 3. If $d \equiv 2 \pmod 3$, then $1 + d \equiv 0 \pmod 3$, that is, a_2 is divisible by 3. Both are impossible, since a_2 and a_3 are relatively prime to n. Hence, $d \equiv 0 \pmod 3$. Since $n - 1 = a_k = 1 + (k-1)d \equiv 1 \pmod 3$. This also cannot be true since $3|n$. This contradiction shows that $d = 1$ or 2.

20. Determine all ordered pairs (m, n) of positive integers such that $\dfrac{n^3+1}{mn-1}$ is an integer.

Solution The relations

$$m^3(n^3+1) = (m^3n^3 - 1) + (m^3+1),$$

$$n^3(m^3+1) = (m^3n^3 - 1) + (n^3+1)$$

show that n^3+1 is divisible by $mn-1$ if and only if m^3+1 is divisible by $mn-1$. Because of this symmetry, we may suppose that $m \leq n$, and that m^3+1 is divisible by $mn-1$. If $m = n$, then

$$\frac{m^3+1}{mn-1} = \frac{n^3+1}{n^2-1} = \frac{n^2-n+1}{n-1} = n + \frac{1}{n-1},$$

and hence $n = 2$. Thus, we get one pair $(2, 2)$.

Since $m^3 + 1 = m(m^2+n) - (mn-1)$, we see that $mn-1$ divides $m(m^2+n)$. But $mn-1$ is relatively prime to m; hence it divides m^2+n. Thus, $m^2+n = k(mn-1)$.

First suppose that $k = 1$, that is, $m^2 + n = mn - 1$. Then $m^2 + 1 = n(m-1)$. So $m > 1$, and $n = m+1+\dfrac{2}{m-1}$. It follows that $m-1 = 1$ or 2. Thus, we get two pairs $(2, 5)$ and $(3, 5)$.

Finally, suppose that $k \geq 2$ and $m < n$. Then $m^2 + n \geq 2(mn - 1)$. Since $m + 1 \leq n$, we get

$$m^2 + 2 \geq n(2m - 1) \geq (m + 1)(2m - 1) = 2m^2 + m - 1.$$

Thus, $m^2 + m \leq 3$, and so $m = 1$. In that case,

$$\frac{m^2 + n}{mn - 1} = \frac{n + 1}{n - 1} = 1 + \frac{2}{n - 1}.$$

It follows that $n - 1 = 1$ or 2. Hence $n = 2$ or 3. We get two more pairs $(1, 2)$ and $(1, 3)$.

If (m, n) is a solution, so is (n, m). Thus, we get 9 solutions, namely,

$$(2, 2), (2, 5), (3, 5), (1, 2), (1, 3), (5, 2), (5, 3), (2, 1), (3, 1).$$

21. Let p be a prime. Show that the congruence $x^2 + x + 3 \equiv 0 \pmod{p}$ is solvable if and only if $x^2 + x + 25 \equiv 0 \pmod{p}$ is solvable.

Solution Since $x(x + 1)$ is even for integers x, we see that $x^2 + x + 3$ and $x^2 + x + 25$ are always odd. So the congruences have no solution when $p = 2$. If $p = 3$, the first is satisfied by $x = 0$ and the second by $x = 1$. So both have solutions when $p = 3$.

Let $p > 3$, and $x = r$ be a root of the first. Then, $9r^2 + 9r + 27 \equiv 0 \pmod{p}$, and hence, $(3r + 1)^2 + (3r + 1) + 25 \equiv 0 \pmod{p}$. So $x = 3r + 1$ is a root of the second.

Suppose, on the other hand, that $x = s$ is a root of the second congruence. Since $p \neq 3$, we see that s, $s+p$ and $s+2p$ are incongruent modulo 3. So one of them is equal to $3r + 1$ for some r. Since all these satisfy the second congruence, we get

$$(3r + 1)^2 + (3r + 1) + 25 \equiv 0 \pmod{p},$$

$$9(r^2 + r + 3) \equiv 0 \pmod{p}.$$

Since $(9, p) = 1$, we see that $x = r$ satisfies the first congruence.

22. Let the sum of reciprocals of integers $1, 2, 3, \ldots, 100$ be written as a/b, where a, b are positive integers which are relatively prime. Prove that ab is not divisible by 5.

Solution Let S_1 denote the sum of reciprocals of numbers which are divisible by 25, S_2 denote the sum of reciprocals of numbers which are divisible by 5 but not by 25, and S_3 be the sum of reciprocals of the remaining numbers. Denote the sum of reciprocals of $5n+1, 5n+2, 5n+3$ and $5n+4$ by u_n. Thus,

$$S_1 = \frac{1}{25} + \frac{1}{50} + \frac{1}{75} + \frac{1}{100} = \frac{1}{12},$$

$$S_2 = \sum_{n=0}^{3} \frac{u_n}{5}, \quad S_3 = \sum_{n=0}^{20} u_n.$$

Now,

$$\frac{1}{5n+1} + \frac{1}{5n+4} = \frac{10n+5}{(5n+1)(5n+4)},$$

$$\frac{1}{5n+2} + \frac{1}{5n+3} = \frac{10n+5}{(5n+2)(5n+3)}.$$

Hence

$$u_n = \frac{(10n+5)(50n^2+50n+10)}{(5n+1)(5n+2)(5n+3)(5n+4)}.$$

The numerator of this is divisible by 25, and the denominator is relatively prime to 5. Thus, $u_n = 25a_n/b_n$, where $\gcd(b_n, 5) = 1$. It follows that $S_2 = 5x_1/y_1$ and $S_3 = 25x_2/y_2$, where y_1, y_2 are relatively prime to 5. Thus, the sum of the reciprocals of numbers from 1 to 100 is equal to

$$\frac{1}{12} + \frac{5x_1}{y_1} + \frac{25x_2}{y_2} = \frac{y_1y_2 + 12(5x_1y_2 + 25x_2y_1)}{12y_1y_2}.$$

The numerator and the denominator are relatively prime to 5. This proves the result.

23. Let p be a prime greater than 3. If the sum of reciprocals of $1, 2, \ldots, p-1$ is m/n, where m and n are integers, prove that $p^2|m$.

Solution Let $f(x) = (x-1)(x-2)\cdots(x-p+1) - (x^{p-1}-1)$. Clearly, this is a polynomial of degree $p-2$. By Fermat's theorem, the congruence $x^{p-1} - 1 \equiv 0 \pmod{p}$ holds for $x = 1, 2, \ldots, p-1$. Since the product is zero for these values of x, we see that $f(x) \equiv 0 \pmod{p}$ for more than $p - 2$ values of x. Hence all its coefficients are divisible by p. Let

$$f(x) = a_0 + a_1 x + a_2 x^2 + \cdots a_{p-2} x^{p-2}.$$

Then, a_1 and a_2 are divisible by p. Note that $2 < p - 2$. Hence,

$$f(p) \equiv a_0 + a_1 p + a_2 p^2 \equiv a_0 + a_1 p \pmod{p^3}.$$

Since

$$a_0 = (-1)^{p-1}(p-1)! + 1, \quad f(p) = (p-1)! - p^{p-1} + 1,$$

we get $a_1 p \equiv 0 \pmod{p^3}$ and hence, $a_1 \equiv 0 \pmod{p^2}$. It can be seen that a_1 is the sum of products of $p-2$ distinct numbers in $\{1, 2, \ldots, p-1\}$. Hence, $a_1/(p-1)!$ is the sum of reciprocals of $1, \ldots, p-1$. Thus,

$$\frac{a_1}{(p-1)!} = \frac{m}{n}, \quad n a_1 = m(p-1)!.$$

Since a_1 is divisible by p^2 and $(p-1)!$ is relatively prime to p, we see that m is divisible by p^2.

24. Show that the equation

$$x^4 + y^4 + z^4 - 2x^2y^2 - 2y^2z^2 - 2z^2x^2 = 24$$

has no solution in integers.

Solution Suppose that there exist integers x, y, z satisfying the equation. Then, $x^4 + y^4 + z^4$ is even. So one or all of x, y, z must be even. If all are even, then each term in the left-hand side of the equation is divisible by 16, while the right-hand side is not divisible by 16. If one is even and the other two are odd, let x be even and y, z be odd. The left-hand side can be written as

$$x^4 + (y^2 - z^2)^2 - 2x^2(y^2 + z^2).$$

Since $y^2 - z^2$ is divisible by 8 and $y^2 + z^2$ is even, this is also divisible by 16. This contradiction proves the result.

25. If x and y are positive integers such that xy divides $x^2 + y^2 - x$, show that x is a perfect square.

Solution There is an integer m such that $x^2 + y^2 - x = mxy$. Let $d = \gcd(x, y)$, $x = x_1 d$ and $y = y_1 d$. Then,

$$x_1^2 d^2 + y_1^2 d^2 - x_1 d = m x_1 y_1 d^2.$$

Cancel d. The result can be written in two ways as

$$(x_1^2 + y_1^2 - m x_1 y_1) d = x_1,$$
$$y_1^2 d = x_1 (m y_1 d - x_1 d + 1).$$

The first shows that $d | x_1$. Since x_1 is prime to y_1, the second shows that $x_1 | d$. Hence, $x_1 = d$ and $x = d^2$.

26. If n is a natural number for which $4^n + 2^n + 1$ is a prime, show that n is a power of 3.

Solution If $n = 1$, then $n = 3^0$. Let $N = 4^n + 2^n + 1$. If n is even, then $N \equiv 1 + 1 + 1 \pmod 3$. So N cannot be prime when n is even. Let n be odd and $n > 1$. If n is not a power of 3, then it must be divisible by a prime $p > 3$. Assume that $n = mp$, where p is prime and $p > 3$. Let $a = 2^m$. Then, $2^n = a^p$, and hence

$$N = 4^n + 2^n + 1 = \frac{2^{3n} - 1}{2^n - 1} = \frac{a^{3p} - 1}{a^p - 1},$$
$$N \left(\frac{a^p - 1}{a - 1} \right) = \frac{a^{3p} - 1}{a^3 - 1} (a^2 + a + 1). \qquad (*)$$

Now, $p = 3k + 1$ or $3k + 2$ for some k. Let

$$u = \frac{a^{3k} - 1}{a - 1} = \frac{a^{3k} - 1}{a^3 - 1} (a^2 + a + 1).$$

So the right side of (*) and u are divisible by $a^2 + a + 1$.

Let $p = 3k + 1$. Then,

$$a^p - 1 = a(a^{3k} - 1) + (a - 1), \qquad \frac{a^p - 1}{a - 1} = au + 1.$$

So we see from (*) that $N(au + 1)$ is divisible by $a^2 + a + 1$, and hence N must be divisible by $a^2 + a + 1$.

Let $p = 3k + 2$. Then,

$$a^p - 1 = a^2(a^{3k} - 1) + (a^2 - 1), \qquad \frac{a^p - 1}{a - 1} = a^2 u + (a + 1).$$

So we see from (*) that $N(a^2u+a+1)$ is divisible by a^2+a+1. Hence, $N(a+1)$ is divisible by a^2+a+1. Since $a^2+a+1-a(a+1)=1$, we see that a^2+a+1 is relatively prime to $a+1$, and hence N is divisible by a^2+a+1.

Thus, we see that N is divisible by a^2+a+1 in both cases. Since $a^2+a+1=2^{2m}+2^m+1<N$, this shows that N cannot be prime. This contradiction shows that n cannot have a prime divisor $p>3$.

27. Find all natural numbers n which are sums of two odd composite natural numbers.

Solution No odd integer can be expressed as the sum of two odd numbers. So consider an even number n. It is of the form $6k$, $6k+2$ or $6k+4$. We have

$$6k = 9+(6k-9), \quad 6k+2 = 35+(6k-33),$$
$$6k+4 = 25+(6k-21).$$

The numbers $6k-9$, $6k-33$ and $6k-21$ are odd and divisible by 3. So they are composite if $6k-33>3$. Hence, the above expressions show that any even $n \geq 42$ can be expressed as a sum of two odd composite numbers.

The odd composite numbers less than 42 are 9, 15, 21, 25, 27, 33, 35, 39. So the even numbers less than 42 and expressible as a sum of two odd composite numbers are 18, 24, 30, 34, 36 and 40.

28. Show that there exists a set S of 2000 distinct positive integers such that each element of S is less than 10^5 and no three of S form an arithmetic progression.

Solution Let S be the set of all positive integers expressible as

$$n = a_0 + a_1\cdot 3 + a_2\cdot 3^2 + \cdots + a_{10}\cdot 3^{10},$$

where $a_i = 0$ or 1 for each i. Since each a_i has two choices, there are $2^{11}-1>2000$ elements in S. The elements are at most

$$1+3+3^2+\cdots+3^{10} = \frac{3^{11}-1}{2} = 88573 < 10^5.$$

Suppose that three elements x, y, z form an arithmetic progression. If y is the middle term, then $x+z=2y$. Let

$$x=\sum_{i=0}^{10} x_i\cdot 3^i, \quad y=\sum_{i=0}^{10} y_i\cdot 3^i, \quad z=\sum_{i=0}^{10} z_i\cdot 3^i.$$

Then

$$\sum_{i=0}^{10}(x_i + z_i)3^i = \sum_{i=0}^{10}(2y_i)3^i.$$

Since $0 \leq x_i + z_i \leq 2$ and $0 \leq 2y_i \leq 2$, the uniqueness of the coefficients in a base-3 representation shows that $x_i + z_i = 2y_i$ for each i. If $y_i = 0$, then $x_i + z_i = 0$, and so $x_i = z_i = 0$. If $y_i = 1$, then $x_i + z_i = 2$, and so $x_i = z_i = 1$. Thus, in either case, $x_i = y_i = z_i$ for all i. This means that $x = y = z$. Hence, no three distinct elements of S can form an arithmetic progression.

29. Note that $12^2 = 144$ ends in two 4's. Determine the length of the longest string of equal non-zero digits in which the square of an integer can end.

Solution The end digits (that is, the digits in the unit's place) of a perfect square can be 1, 4, 9, 6 or 5 only. If the end digits of n are 11, 99 or 55, then n is of the form $4k + 3$; and if they are 66, then $n = 4k + 2$. But the square of any integer is of the form $4k$ or $4k + 1$. Thus, if a square ends in a string of at least two equal non-zero digits, then that digit must be 4. Suppose that a^2 ends in 4444. Then, a is even, and

$$a^2 = 10^4m + 4444 \equiv 12 \pmod{16}.$$

But $(4k)^2 \equiv 0 \pmod{16}$ and $(4k + 2) \equiv 4 \pmod{16}$. So $a^2 \not\equiv 12 \pmod{16}$. Hence a square cannot end in 4444. Since $38^2 = 1444$, the desired length of equal non-zero digits is 3.

30. For an integer $k \geq 2$, define $f : \mathbf{N} \longrightarrow \mathbf{N}$ by

$$f(n) = \left[\left(n + n^{1/k}\right)^{1/k}\right] + n,$$

where $[x]$ denotes the integral part of x. Prove that the range of f omits exactly those natural numbers which are kth powers.

Solution First suppose that a is not a kth power. Then, $m^k < a < (m + 1)^k$ for some positive integer m. We have $(m - 1)^2 \leq m^2 - m$. In general, as can be seen by induction on k,

$$(m - 1)^k \leq m^k - m, \quad \text{for all } k \geq 2. \tag{*}$$

So $(m - 1)^k < a - m$. Let $n = a - m$. Then, $m - 1 < n^{1/k} < a^{1/k}$. Further, $m^k \leq a - 1$ as $m^k < a$, and $(m+1)^k \geq a+1$ as $a < (m+1)^k$.

So

$$m^k \le a-1 = n+m-1 < n+n^{1/k} < (a-m)+a^{1/k}$$
$$< (a-m)+(m+1) = a+1 \le (m+1)^k.$$

It follows that $m < \left(n+n^{1/k}\right)^{1/k} < m+1$, and $f(n) = m+n = a$. Thus, a belongs to the range of f.

Next suppose that a is a kth power. Let $a = m^k$, where $m > 1$, and $n = m^k - m$. Then,

$$n^{1/k} < m, \quad n+n^{1/k} < m^k, \quad \left(n+n^{1/k}\right)^{1/k} < m.$$

Consequently, $\left[\left(n+n^{1/k}\right)^{1/k}\right] \le m-1$. Hence, $f(n) \le m-1+m^k-m = a-1$.

Further, $n+1 > m^k - m \ge (m-1)^k$ from (*), and hence,

$$(n+1)^{1/k} > m-1, \quad n+1+(n+1)^{1/k} > m^k.$$

As before, $f(n+1) \ge m+m^k-m+1 = a+1$. It is clear that $f(n)$ is a strictly increasing function of n. Since $a = m^k$ lies between two successive values of $f(n)$, we see that the equality $f(n) = a$ can never be true. Further, $f(1) = 2$, and so there is no n with $f(n) = 1$. So a does not belong to the range of f if it is a kth power.

Set VII

1. Let n be a positive integer. Prove that if $2+2\sqrt{28n^2+1}$ is an integer, then it is a square.

   Solution Let $N = 2+2\sqrt{28n^2+1}$ be an integer. Then, $\sqrt{28n^2+1}$ must be an integer m, and $m^2 = 28n^2+1$. Hence, m is odd, and $7n^2 = (m^2-1)/4 = (m-1)/2 \cdot (m+1)/2$. Note that $(m-1)/2$ and $(m+1)/2$ are relatively prime, and one of them is to be divisible by 7. If the product of two coprime factors is a perfect square, then both factors must be perfect squares. It follows that either (i) $(m-1)/2 = 7a^2$ and $(m+1)/2 = b^2$ or (ii) $(m-1)/2 = a^2$ and $(m+1)/2 = 7b^2$. In case (i), $N = 2(1+m) = 4b^2$, a perfect square. In case (ii), $a^2 = (m+1)/2 - 1 = 7b^2 - 1 \equiv 6 \pmod 7$. But a square is congruent modulo 7 to 0, 1, 4 or 2. So case (ii) is impossible.

2. Find all pairs (n, k) of positive integers such that the congruence

$$a^k \equiv 0 \text{ or } 1 \pmod{n^k}$$

holds for every integer a.

Solution If $n = 1$ then $n^k = 1$, and any value of k satisfies the condition. Suppose that the condition is satisfied for some $n > 1$ and $k \geq 1$. If k is odd and $a = n^k - 1$, then $a^k \equiv -1 \pmod{n^k}$. So the condition is satisfied only when $-1 \equiv 1 \pmod{n^k}$. This implies that $n^k | 2$, that is, $n^k = 2$, $n = 2$ and $k = 1$.

If $k = 2r$ and $a = n^r + 1$, then

$$a^k = (1 + n^r)^{2r} \equiv 1 + (2r)n^r \pmod{n^{2r}}.$$

So the condition shows that

$$1 + (2r)n^r \equiv 0 \text{ or } 1 \pmod{n^{2r}}.$$

If $1 + (2r)n^r$ is divisible by n^{2r}, then it must be divisible by n^r. Since this is not possible, we see that

$$1 + (2r)n^r \equiv 1 \pmod{n^{2r}}, \quad 2r \equiv 0 \pmod{n^r}.$$

Thus, $n^r | 2r$ and hence, $n^r \leq 2r$. But $n \geq 2$, and $2^r > 2r$ for all $r \geq 3$, as can be seen by induction on r. It follows that $r < 3$. If $r = 1$, then we get $n | 2$, and if $r = 2$, then $n^2 | 4$. In either case, we must have $n = 2$.

It remains to consider the congruences

$$a^k \equiv 0 \text{ or } 1 \pmod{2^k} \quad \text{for all integers } a,$$

when $k = 1, 2$ or 4. If a is even, then $a^k \equiv 0 \pmod{2^k}$. If $a = 2n + 1$, then $a \equiv 1 \pmod 2$, $a^2 \equiv 1 \pmod 4$, and

$$a^4 \equiv 1 + 4(2n) + 6(4n^2) \equiv 1 + 8n(n+1) \equiv 1 \pmod{16}.$$

Hence, these are true. Thus, the desired pairs (n, k) are $(1, k)$, for any k, and (2, 1), (2, 2) and (2, 4).

3. Determine all integral solutions of the equation $x^2 + y^2 + z^2 = x^2 y^2$.

Solution Clearly, $x = y = z = 0$ is a solution. Suppose that (x, y, z) is any solution in which at least one is non-zero. If x or y is odd, then $x^2 - 1$ or $y^2 - 1$ is divisible by 4, and hence $(x^2 - 1)(y^2 - 1) \equiv 0$

(mod 4). But $(x^2-1)(y^2-1) = z^2+1$ and $z^2+1 \equiv 1$ or $2 \pmod 4$. This contradiction shows that x and y are even. In that case, z too must be even, as can be seen from the given equation.

Since at least one of x, y, z is non-zero, we can find an integer r such that 2^r is the highest power of 2 that divides x, y and z. Let $x = 2^r x_1$, $y = 2^r y_1$, $z = 2^r z_1$. Then one of x_1, y_1, z_1 must be odd. The given equation reduces to

$$x_1^2 + y_1^2 + z_1^2 = 2^{2r} x_1^2 y_1^2.$$

Since $r \geq 1$, the right-hand side of this is congruent to 0 modulo 4. Since the square of an integer is congruent to 0 or 1 modulo 4, the left-hand side is congruent to $1+1+1$, $1+1+0$ or $1+0+0$. This gives a contradiction. So $x = y = z = 0$ is the only solution.

4. Determine all integers k such that the equation $x^2 + y^2 + z^2 = kxyz$ has a non-trivial solution in integers.

Solution If (x, y, z) is a solution for some k, then $(-x, -y, -z)$ is a solution for $-k$. Since $k = 0$ gives only the trivial solution $x = y = z = 0$, we need consider only positive integral values for k.

If $k = 1$, then $x = y = z = 3$ gives a solution; and if $k = 3$, then $x = y = z = 1$ is a solution. Let k be a positive integer other than 1 and 3. We shall prove that the equation has only the trivial solution $x = y = z = 0$.

Suppose that the equation has a non-trivial solution in integers x, y, z. If at least one of x, y, z is zero, then the remaining are also zero. So let x, y, z be positive. We can choose (x, y, z) for which the sum $x+y+z$ has the least value. Because of the symmetry, we may suppose that $x \leq y \leq z$. Note that z is a root of the quadratic equation

$$t^2 - (kxy)t + x^2 + y^2 = 0 \qquad (*)$$

in t. Let z_1 be its second root. Then,

$$\begin{aligned}(y-z)(y-z_1) &= y^2 - (z+z_1)y + zz_1 = y^2 - kxy^2 + x^2 + y^2 \\ &\leq y^2 - kxy^2 + y^2 + y^2 = y^2(3-kx).\end{aligned}$$

Clearly, $kx \geq 4$ for all $x \geq 1$ if $k \geq 4$, and also for $x \geq 2$ if $k = 2$. If $k = 2$ and $x = 1$, then the equation becomes $1 + (y-z)^2 = 0$, which has no solution. Thus, if $k \neq 1$ and $k \neq 3$, then we must have

$kx \geq 4$, and hence $(y - z)(y - z_1) < 0$. Since $y \leq z$, this implies that $z_1 < y < z$. Since $z_1 = kxy - z$ is an integer satisfying (*), we see that (x, y, z_1) is a solution in positive integers of the equation. But $x+y+z_1 < x+y+z$. This contradicts the choice of (x, y, z). It follows that the equation cannot have a non-trivial solution unless $k = \pm 1, \pm 3$.

If $k = 1$ or 3, then the equation has infinitely many solutions. If (x, y, z) is a solution with $x \leq y \leq z$ and $kx = 3$, let $y_1 = 3z - y$. Then (x, z, y_1) is a new solution with $x \leq z < y_1$. If $k = 1$, we can start with $(3, 3, 3)$ and get, successively, $(3, 3, 6)$, $(3, 6, 15)$, $(3, 15, 39)$, and so on. If $k = 3$, we can start with $(1, 1, 1)$ giving $(1, 1, 2)$, $(1, 2, 5)$, $(1, 5, 13)$, and so on.

5. Let r be an integer greater than 1 and $2 \leq n \leq (2^r - 1)^2$. If $x_1, \ldots, x_n$ are integers such that

$$x_1^r + x_2^r + \cdots + x_n^r = x_1^r x_2^r \cdots x_n^r, \quad x_i \geq 0 \text{ for all } i.$$

Prove that $x_1 = x_2 = \cdots = x_n = 0$.

Solution We may suppose that $x_1 \leq x_2 \leq \cdots \leq x_n$. If $x_1 = 0$, then from the given equation we get $x_2 = \cdots = x_n = 0$. So let $x_1 \geq 1$. If $n = 2$, then $(x_1^r - 1)(x_2^r - 1) = 1$ from the given equation. This implies that $x_1^r - 1 = x_2^r - 1 = 1$, and x_1, x_2 are irrational. So let $n > 2$.

First suppose that $x_{n-1} \geq 2$. If, further, $x_{n-2} \geq 2$, then $x_{n-2}x_{n-1} \geq 4$, and hence,

$$x_1^r x_2^r \cdots x_n^r \geq 4^r x_n^r > n x_n^r \geq x_1^r + x_2^r + \cdots + x_n^r,$$

since $4^r > (2^r - 1)^2 \geq n$. This contradicts the given equality.

Thus, we see that $x_{n-2} = 1$, which implies that $x_i = 1$ for $i \leq n - 2$. Consequently,

$$(n - 2) + 2x_n^r \geq x_1^r + \cdots + x_{n-1}^r + x_n^r = x_1^r \cdots x_{n-1}^r x_n^r \geq 2^r x_n^r,$$
$$(2^r - 2)x_n^r \leq n - 2 \leq (2^r - 1)^2 - 2 = 2^r(2^r - 2) - 1.$$

This implies that $x_n < 2$, and so $x_n < x_{n-1}$. The contradiction shows that $x_{n-1} = 1$. Consequently, $x_i = 1$ for $i \leq n - 1$, and hence, the given equation becomes $n - 1 + x_n^r = x_n^r$, which is not possible.

Note that if $n = (2^r - 1)^2 + 1$, then we have a non-trivial solution, namely, $x_1 = \cdots = x_{n-2} = 1$, $x_{n-1} = x_n = 2$.

6. Show that the equation

$$x_1^4 + x_2^4 + \cdots + x_{14}^4 = 1599$$

has no solution in integers.

Solution Consider the least non-negative residues modulo 16. The right-hand side of the equation is congruent to 15 modulo 16. Let $x_i^4 \equiv a_i \pmod{16}$, where $0 \le a_i < 16$. Then the left-hand side is congruent to $a_1 + \cdots + a_{14}$. Since $(2n)^4 \equiv 0 \pmod{16}$, and

$$(2n+1)^4 \equiv 1 + 4(2n) + 6(2n)^2 \equiv 1 + 8n(n+1) \equiv 1 \pmod{16},$$

we see that $a_i = 0$ or 1 for each i. Hence, $0 \le a_1 + \cdots + a_{14} \le 14$. So the equality can never happen.

7. Let n be a positive integer. Determine whether or not there are distinct positive integers $x_1, \ldots, x_n$ such that

$$x_1^2 + x_2^2 + \cdots + x_n^2 = y^3 \quad \text{and} \quad x_1^3 + x_2^3 + \cdots + x_n^3 = z^2,$$

simultaneously, for integers y, z.

Solution We shall see that there are infinitely many such integers $x_1, \ldots, x_n$. Let

$$1^2 + 2^2 + \cdots + n^2 = a, \quad 1^3 + 2^3 + \cdots + n^3 = b,$$

and $x_i = a^r b^s\, i$, for $i = 1, 2, \ldots, n$, where r and s are to be chosen. Then the x_i are distinct, and

$$x_1^2 + \cdots + x_n^2 = (a^r b^s)^2 a = a^{2r+1} b^{2s},$$
$$x_1^3 + \cdots + x_n^3 = (a^r b^s)^3 b = a^{3r} b^{3s+1}.$$

It is enough if we choose r, s such that $2r + 1$ and $2s$ are divisible by 3, and $3r$ and $3s + 1$ are divisible by 2. This will be true if r is of the form $6m + 4$ and s is of the form $6k + 3$. Thus, we can choose x_i in an infinite number of ways.

8. An integer n will be called *good* if we can write

$$n = a_1 + a_2 + \cdots + a_k,$$

where $a_1, a_2, \ldots, a_k$ are positive integers (not necessarily distinct) such that the sum of their reciprocals is 1. Given the information that the integers 33 through 73 are good, prove that every $n \ge 33$ is good.

Solution Suppose that n is good. Then there are positive integers $a_1, \ldots, a_k$ such that

$$a_1 + a_2 + \cdots + a_k = n, \qquad \frac{1}{a_1} + \frac{1}{a_2} + \cdots + \frac{1}{a_k} = 1.$$

Then,

$$\frac{1}{4} + \frac{1}{4} + \frac{1}{2a_1} + \cdots + \frac{1}{2a_k} = 1, \qquad \frac{1}{3} + \frac{1}{6} + \frac{1}{2a_1} + \cdots + \frac{1}{2a_k} = 1.$$

The sum of the denominators in the first equality is $2n + 8$, and that in the second is $2n + 9$. This shows that if n is good, so are $2n + 8$ and $2n + 9$. Since $n = 33$ is good, $2n + 8 = 74$ and $2n + 9 = 75$ are good. Thus, the numbers from 34 to 75 are good.

Now consider the following statement: All the integers $n, n + 1, \ldots, 2n + 7$ are good. This is true when $n = 33$ and $n = 34$. Assume that it is true for some n. Since $2(n + 1) + 7 = 2n + 9$, we see that all the integers $n + 1, n + 2, \ldots, 2(n + 1) + 7$ are good, that is, the statement is true for $n + 1$. This proves the result by induction.

9. Prove that the cube roots of three distinct prime numbers cannot be three terms (not necessarily consecutive) of an arithmetic progression.

Solution Let the nth, mth and kth terms of an arithmetic progression be x, y and z, respectively, where $n < m < k$. Then, $y - x = (m - n)d$ and $z - y = (k - m)d$, where d is the common difference. This implies that

$$\begin{aligned} (y - x)(k - m) &= (z - y)(m - n), \\ y(k - n) &= x(k - m) + z(m - n). \end{aligned}$$

Suppose that $x = \sqrt[3]{p_1}$, $y = \sqrt[3]{p_2}$ and $z = \sqrt[3]{p_3}$, where p_1, p_2, p_3 are distinct primes. It is known that if $a + b + c = 0$, then $a^3 + b^3 + c^3 = 3abc$. Using this, we get

$$\begin{aligned} x^3(k - m)^3 &- y^3(k - n)^3 + z^3(m - n)^3 \\ &= -3xyz(k - m)(k - n)(m - n). \end{aligned}$$

Since x^3, y^3, z^3 are integers, this implies that xyz is a rational number, that is, $\sqrt[3]{p_1 p_2 p_3}$ is rational. This is not true. Hence, x, y, z cannot be cube roots of primes.

10. Show that the cube roots of three distinct prime numbers cannot be three terms (not necessarily consecutive) of a geometric nor of a harmonic progression.

Solution Let, as in the above problem, x, y and z be the nth, mth and kth terms of (i) a geometric, or (ii) a harmonic, progression. First consider case (i). We have $y/x = r^{m-n}$ and $z/y = r^{k-m}$, where r is the common ratio of the progression. This implies that

$$\left(\frac{y}{x}\right)^{k-m} = \left(\frac{z}{y}\right)^{m-n}, \quad y^{k-n} = x^{k-m} \cdot z^{m-n} \qquad (*)$$

Suppose that $x = \sqrt[3]{p_1}$, $y = \sqrt[3]{p_2}$ and $z = \sqrt[3]{p_3}$, where p_1, p_2, p_3 are distinct primes. Cubing the second equality in (*), we get

$$p_2^{k-n} = p_1^{k-m} \cdot p_3^{m-n}.$$

This contradicts the unique factorisation theorem.

Next consider case (ii). We have $1/y - 1/x = (m-n)d$ and $1/z - 1/y = (k-m)d$, where d is the common difference of the arithmetic progression of the reciprocals. It follows that

$$(k-n)\frac{1}{y} = (k-m)\frac{1}{x} + (m-n)\frac{1}{z},$$

$$(k-n)zx = (k-m)yz + (m-n)xy.$$

As in the case of arithmetic progressions, we get

$$\begin{aligned}(k-m)^3(yz)^3 - (k-n)^3(zx)^3 + (m-n)^3(xy)^3 \\ = -3x^2y^2z^2(k-m)(k-n)(m-n).\end{aligned}$$

This implies that $p_1^{2/3}p_2^{2/3}p_3^{2/3}$ is rational, which is not true, and this completes the proof.

11. Prove that, for every positive integer n there is a finite set S_n of positive integers such that it has n elements and the geometric mean of every subset of S_n is an integer. Is there an infinite set S of positive integers such that the geometric mean of any finite subset of S is an integer?

Solution Choose n distinct positive integers $a_1, a_1, \ldots, a_n$, and let

$$S_n = \{a_1^{n!}, a_2^{n!}, \ldots, a_n^{n!}\}.$$

Then the geometric mean of any non-empty subset of S_n is an integer.

The answer to the second part is 'No'. Assume that there is an infinite set S with the desired property. Let a and b be two distinct elements in S. By the unique factorisation theorem, $a/b = p_1^{r_1} p_2^{r_2} \cdots p_k^{r_k}$, where $p_1, \ldots, p_k$ are distinct primes and $r_1, \ldots, r_k$ are non-zero integers either positive or negative. The nth root of a/b is a rational number if and only if each r_i is divisible by n. This can be true for only a finite number of values of n. Choose n for which this is not true. Then choose $n-1$ elements $x_1, \ldots, x_{n-1}$ from S. If the geometric means of the sets $\{a, x_1, \ldots, x_{n-1}\}$ and $\{b, x_1, \ldots, x_{n-1}\}$ are integers, then their ratio is rational. But the ratio is $(a/b)^{1/n}$. This contradicts the choice of n.

12. Find the least value of $n > 1$ such that the root-mean-square of the first n natural numbers is an integer.

Solution By root-mean-square we mean the square root of the arithmetic mean of squares. So if m is the root-mean-square of $1, 2, \ldots, n$, then

$$m^2 = \frac{1^2 + 2^2 + \cdots + n^2}{n} = \frac{n(n+1)(2n+1)}{6n} = \frac{(n+1)(2n+1)}{6}.$$

We have to choose n so that m is an integer. Hence, $6|(n+1)(2n+1)$. Since $2n+1$ is odd, we see that $n+1$ is even, and 3 divides $(n+1)/2$ or $2n+1$.

In the first case, $(n+1)/6$ and $2n+1$ are relatively prime integers, and their product is a square. This implies that $(n+1) = 6a^2$ and $2n+1 = b^2$, and so

$$12a^2 - b^2 = 1, \quad b^2 \equiv 3 \pmod 4$$

which is not possible. Hence, $3|2n+1$. Then, $(n+1)/2$ and $(2n+1)/3$ are relatively prime, and their product is a square. This implies that $n+1 = 2a^2$ and $2n+1 = 3b^2$. Thus, $4a^2 - 3b^2 = 1$ and $(2a-1)(2a+1) = 3b^2$. As above, we see that (i) $2a-1 = 3x^2$, $2a+1 = y^2$ or (ii) $2a+1 = 3x^2$, $2a-1 = y^2$, where x and y are integers. In case (i), $y^2 = 3x^2 + 2 \equiv 2 \pmod 3$, which is not possible. In case (ii), $y^2 = 3x^2 - 2$. This is satisfied by $x = 1, y = 2$ and by $x = 3, y = 5$. Note that $a = (3x^2 - 1)/2$ and $n = 2a^2 - 1$. If $x = 1$, then $a = 1$, $n = 1$. If $x = 3$, then $a = 13$, $n = 337$. The least value of $n > 1$ is 337. The next feasible value of x is 11 with $y = 19$, giving $a = 181$.

13. Find all values of n such that $1 < n < 100$ and the standard deviation of the first n natural numbers is an integer.

Solution The standard deviation of $x_1, x_2, \ldots, x_n$ is the square root of the number

$$\frac{1}{n}\sum_{i=1}^{n} x_i^2 - \left(\frac{1}{n}\sum_{i=1}^{n} x_i\right)^2.$$

So if s is the standard deviation of $1, 2, \ldots, n$, then

$$s^2 = \frac{n(n+1)(2n+1)}{6n} - \left(\frac{n(n+1)}{2n}\right)^2 = \frac{n^2-1}{12}.$$

We have to choose n so that s is an integer.

It can be seen that $n^2 - 1$ is divisible by 12 if and only if n is of the form $n = 6k \pm 1$, in which case

$$s^2 = \frac{n^2-1}{12} = \frac{36k^2 \pm 12k}{12} = k(3k \pm 1).$$

Hence k and $3k \pm 1$ must be perfect squares. But $3k - 1 \equiv 2 \pmod 3$, and so $3k-1$ cannot be a square. It follows that $n = 6k+1$ with $k = x^2$ and $3k + 1 = y^2$. This is satisfied by $x = 1, y = 2$ giving $n = 7$, $s = 2$. The next possible value of x is 4, giving $n = 97$, $s = 28$. There are no other values between 1 and 100.

14. (a) Show that there exists an odd positive integer k such that $2^n + k$ is composite for all positive integers n.

(b) Show that there exists a positive integer k such that $k \cdot 2^n + 1$ is composite for all positive integers n.

Solution (a) Since $2^2 \equiv 1 \pmod 3$, we first observe that $2^n + k \equiv 0 \pmod 3$ if n is divisible by 2 and $k + 1 \equiv 0 \pmod 3$. Similarly, since $2^3 \equiv 1 \pmod 7$, $2^n + k \equiv 0 \pmod 7$ if n is divisible by 3 and $k + 1 \equiv 0 \pmod 7$. Since $k + 1$ is to be even, we let $k + 1 \equiv 0 \pmod{42}$. Then, $2^n + k$ is composite whenever n is divisible by 2 or 3. If $n = 24N + r$, where $1 \le r \le 24$, then $2^n + k$ is composite except possibly when r equals 1, 5, 7, 11, 13, 17, 19 or 23.

Similarly, since $2^4 \equiv 1 \pmod 5$, we see that if $r = 1, 5, 13$ or 17, then $2^r \equiv 2 \pmod 5$. It follows that $2^{24N+r} + k$ is divisible by 5 if $k + 2 \equiv 0 \pmod 5$.

Since $2^8 \equiv 1 \pmod{17}$, we see that if $r = 11$ or 19, then $2^r \equiv 2^3 \pmod{17}$. It follows that $2^{24N+r} + k$ is divisible by 17 if $k + 8 \equiv 0 \pmod{17}$.

Again, since $2^{12} \equiv 1 \pmod{13}$, we see that $2^{24N+23} + k$ is divisible by 13 if $k + 2^{23} \equiv 0 \pmod{13}$.

It remains to consider one more value of r, namely, 7. We have

$$2^{24} - 1 = (2^{12} - 1)(2^{12} + 1) = 4095 \cdot 4097 = 4095 \cdot 17 \cdot 241,$$

and hence, $2^{24} \equiv 1 \pmod{241}$. Thus, $2^{24N+7} + k$ is divisible by 241 if $k + 2^7 \equiv 0 \pmod{241}$.

We have reduced the problem to one of determining k satisfying all the following congruences:

$$k + 1 \equiv 0 \pmod{42}, \quad k + 2 \equiv 0 \pmod{5},$$
$$k + 8 \equiv 0 \pmod{17}, \quad k + 2^{23} \equiv 0 \pmod{13},$$
$$k + 2^7 \equiv 0 \pmod{241}.$$

Since the moduli are pair-wise coprime, this simultaneous system has a solution by the Chinese remainder theorem.

We have proved that there is an odd number k such that $2^n + k$ is divisible by one of the primes 3, 5, 7, 13, 17, 241. If m is the product of 42, 5, 17, 13 and 241, then the same is true when k is replaced by $k + m$. It follows that we can find an infinite number of odd numbers with the desired property. We can see that $k = 8\,871\,323$ is a solution, and $m = 11\,184\,810$.

(b) It is possible to find odd positive integers k such that, for each $n \geq 1$, $2^n + k$ is divisible by one of the primes 3, 5, 7, 13, 17, 241. Clearly, k is not divisible by any of these primes. Let m be their product. Then, k is relatively prime to m, and hence $kk_1 \equiv 1 \pmod{m}$ for some positive integer k_1. If $2^n + k$ be divisible by p, where $p = 3, 5, 7, 13, 17$ or 241, then

$$k(k_1 \cdot 2^n + 1) = (kk_1 - 1)2^n + 2^n + k \equiv 0 \pmod{p},$$

since $kk_1 - 1$ is divisible by m, and hence by p. It follows that $k_1 \cdot 2^n + 1 \equiv 0 \pmod{p}$. It is clear that if k_1 has this property, so does $k_1 + m$. Hence, we can find an infinite number of such values for k_1.

15. Do there exist 14 consecutive positive integers each of which is divisible by one or more primes p from the interval $2 \leq p \leq 11$?

    Solution The answer is 'No'. Consider any set of 14 consecutive integers. Of them, 7 are odd. Suppose that $n, n+2, \ldots, n+12$ are the odd numbers. Let S be the set of those odd numbers which are not divisible by 3. No element in S is divisible by 2 or 3. Let A_p be the set of elements in S which are divisible by the odd prime p.

    Suppose that n is divisible by 3. Then, $n+6$ and $n+12$ are divisible by 3, and $S = \{n+2, n+4, n+8, n+10\}$. Any two elements in A_p differ by $2p$. If $|A|$ denotes the number of elements in a set A, then $|A_{11}| = 1$, $|A_7| = 1$ and $|A_5| = 1$. Thus, $A_5 \cup A_7 \cup A_{11}$ contains at most $1+1+1$ elements, while $|S| = 4$. Hence, S must contain an element not divisible by any prime p with $2 \leq p \leq 11$.

    Next suppose that n is not divisible by 3. Then, only two of the odd numbers can be divisible by 3, and the remaining five are in S. Since $|A_5| = 1$ or 2 and $|A_7| = |A_{11}| = 1$, we see that $A_5 \cup A_7 \cup A_{11}$ contains at most $2+1+1$ elements, while $|S| = 5$. Again, S must contain an element not divisible by any prime p with $2 \leq p \leq 11$.

16. Do there exist 21 consecutive positive integers each of which is divisible by one or more primes p from the interval $2 \leq p \leq 13$?

    Solution The answer is 'Yes'. Let n be divisible by $2 \cdot 3 \cdot 5 \cdot 7 = 210$. Consider the 21 consecutive integers $n-10, n-9, \ldots, n+10$. Of these, all but $n-1$ and $n+1$ are divisible by a prime $p \leq 7$. Hence, it is enough if we choose $n = 210k$ such that $n - 1 \equiv 0 \pmod{11}$ and $n + 1 \equiv 0 \pmod{13}$. Since $210k \equiv k \pmod{11}$ and $210k \equiv 2k \pmod{13}$, we have to choose k such that $k \equiv 1 \pmod{11}$ and $2k \equiv 12 \pmod{13}$. This is satisfied by $k = 45$, and hence $n = 9450$. The corresponding consecutive integers are $9440, 9441, \ldots, 9460$.

17. The nth triangular number t_n is given by $t_n = n(n+1)/2$. Prove that

    (a) n is an even positive integer such that t_n is a perfect square if and only if there exist integers x and y satisfying the equation $x^2 - 2y^2 = 1$ and $n = 2y^2$;

    (b) n is an odd positive integer such that t_n is a perfect square if and only if there exist integers x and y satisfying the equation $x^2 - 2y^2 = -1$ and $n = 2y^2 - 1$;

(c) there exist an infinite number of even integers n, as well as odd integers n, such that t_n is a square.

Solution (a) Suppose that n is even and $t_n = n(n+1)/2$ is the square of an integer. Then, $n/2$ and $n+1$ are relatively prime integers whose product is a perfect square. So both are perfect squares; that is, $n+1 = x^2$ and $n/2 = y^2$ for some integers x, y. Thus, $x^2 - 2y^2 = 1$. On the other hand, if x, y are positive integers such that $x^2 - 2y^2 = 1$ and $n = 2y^2$, then $t_n = y^2x^2$, a perfect square.

(b) Similarly, if n is odd and t_n is a square, then there exist integers x, y such that $n = x^2$ and $(n+1)/2 = y^2$. Thus, $x^2 - 2y^2 = -1$. If x, y are like this and $n = 2y^2 - 1$, then $t_n = x^2y^2$.

(c) Given positive integers a and b, let $x = 3a + 4b$ and $y = 2a + 3b$. Then, $x^2 - 2y^2 = a^2 - 2b^2$. So starting from the solution $(3, 2)$, we can find any number of solutions of the equation $x^2 - 2y^2 = 1$; and starting from $(1, 1)$, we get any number of solutions of $x^2 - 2y^2 = -1$.

18. Determine all positive integers n such that $n!$ is divisible by the sum $1 + 2 + \cdots + n$.

Solution Let $s_n = 1 + 2 + \cdots + n = n(n+1)/2$. We shall prove that $n!$ is divisible by s_n if and only if $n + 1 = 2$ or a composite number.

If $n = 1$, then $s_n = 1$, and if $n = 3$, then $n! = 6 = s_n$. So $n!$ is divisible by s_n when $n = 1$ or 3.

Next, let n be such that $n+1$ is composite and $n \neq 3$. Then $n > 4$, and $n + 1 = ab$, where $1 < a \leq b < n$. If $a < b$, then $n!$ is divisible by nab and $nab = 2s_n$. If $a = b$, then $n + 1 = a^2$. Since $n > 4$, we have $a > 2$, and hence,

$$2a + 2 < 3a \leq a^2 = n + 1, \quad 1 < a < 2a < n - 1.$$

This shows that $n!$ is divisible by $na(2a) = 4s_n$. Thus, $n!$ is divisible by s_n in either case.

In the remaining case, $n = p - 1$, where p is an odd prime. Then, n is even and $s_n = (n/2)p$ is a multiple of p, while $n! = (p-1)!$ is not divisible by p. So $n!$ is not divisible by s_n in this case.

19. If a and b are positive integers such that $a|b^2$, $b^2|a^3$, $a^3|b^4$, $b^4|a^5$, ..., prove that $a = b$.

Solution It is given that $a^{2n-1}|b^{2n}$ and $b^{2n}|a^{2n+1}$. Thus,

$$u_n = \frac{a^{2n+1}}{b^{2n}} \quad \text{and} \quad v_n = \frac{b^{2n}}{a^{2n-1}}$$

are positive integers for all n. Now,

$$u_{n+1} = \frac{a^{2n+3}}{b^{2n+2}} = \frac{a^2}{b^2}u_n, \quad v_{n+1} = \frac{b^{2n+2}}{a^{2n+1}} = \frac{b^2}{a^2}v_n.$$

If $a < b$, then $u_{n+1} < u_n$ for all n. But we cannot have a strictly decreasing infinite sequence $\{u_n\}$ of positive integers. If $a > b$, then $v_{n+1} < v_n$ for all n, and we get a similar contradiction. Hence $a = b$.

20. Prove that if $n > 1$, then $s_n = \sum_{j=1}^{n} 1/j$ is not an integer.

Solution Let $2^k \leq n < 2^{k+1}$, and P be the product of all odd integers in the set $S = \{1, 2, \ldots, n\}$. For $j \in S$, we have $j = 2^r a$, where $r \geq 0$ and a is odd. Then, $2^{k+1} > n \geq j = 2^r a \geq 2^r$, and so $r < k+1$. Moreover, if $a > 1$, then $n \geq 2^r \cdot 3 > 2^{r+1}$, and so $r+1 < k+1$. Hence, if $r = k$, then $a = 1$ in which case, $j = 2^k$. This shows that if $j \in S$, $j \neq 2^k$ and $j = 2^r a$, then $r < k$ and so $2^k(P/j) = 2^{k-r}(P/a)$ is an even integer. Thus, $2^k P s_n = P + \sum_{j \neq 2^k} 2^k(P/j)$ is an odd integer. But if s_n is an integer, then $2^k P s_n$ is even. So s_n cannot be an integer.

21. Prove that if $n > m \geq 1$, then $s_n = \sum_{j=m}^{n} 1/j$ is not an integer.

Solution We can use the same procedure. Let $2^k \leq n < 2^{k+1}$, and P be the product of all odd integers not exceeding n. If $j = 2^r a$, where $r \geq 0$ and a is odd, then $r \leq k$. Suppose that $m \geq 2^k$. Then

$$s_n < \frac{n - m + 1}{m} \leq \frac{2^{k+1} - m}{m} \leq \frac{2^k}{m} \leq 1.$$

Thus, s_n cannot be an integer if $m \geq 2^k$. So suppose that $m < 2^k$. Then, one of the elements in the set $S = \{m, m+1, \ldots, n\}$ is 2^k. Exactly as in the previous problem, we see that if $j \in S$ and $j = 2^r a$, with a odd, then $r = k$ only when $j = 2^k$. This shows that if $j \neq 2^k$ and $j = 2^r a$, then $2^k(P/j) = 2^{k-r}(P/a)$ is an even integer. Thus, $2^k P s_n = P + \sum_{j \neq 2^k} 2^k(P/j)$ is an odd integer. Hence, s_n cannot be an integer, as before.

22. Let d be a positive integer, and

$$s_n = \frac{1}{1+d} + \frac{1}{1+2d} + \cdots + \frac{1}{1+nd}.$$

Assuming Bertrand's theorem, which states that given an integer $n > 1$, we can find a prime p such that $n \le p < 2n$, or otherwise, prove that s_n is not an integer for any $n \ge 1$.

Solution Assuming that s_n is an integer for some n, we shall arrive at a contradiction. Let $x_i = \dfrac{1}{1+id}$ so that $s_n = x_1 + x_2 + \cdots + x_n$. Suppose that $2^k \le n < 2^{k+1}$.

We shall first consider the case in which $k \ge 6$, that is, $n \ge 64$. It can be verified by induction that $2^k \ge 6(k+1)$ if $k \ge 6$. Now,

$$s_n \le \sum_{i<2^{k+1}} x_i = x_1 + (x_2 + x_3) + (x_4 + \cdots + x_7) + \cdots +$$

$$+ \left(x_{2^k} + \cdots + x_{2^{k+1}-1}\right).$$

In each group, the first term is the largest. Hence

$$s_n < x_1 + 2x_2 + 4x_4 + \cdots + 2^k x_{2^k} < \frac{k+1}{d}, \qquad (*)$$

as $ix_i < 1/d$ for all i. Since s_n is assumed to be an integer, we must have

$$d \le ds_n < k+1 \le \frac{2^k}{6} \le \frac{n}{6}.$$

Hence, $n > 6d$.

By Bertrand's theorem, we can find a prime p such that $n/2 \le p \le n$. So $p > 3d$, which implies that $(p, d) = 1$. Let

$$u_i = \frac{1}{x_i} = 1 + id.$$

Then, $u_r \equiv 0 \pmod{p}$ for some r with $1 \le r \le p-1$. First suppose that there is only one such r with $1 \le r \le n$. If $u_r = ap$, then

$$s_n - \frac{1}{ap} = s_n - \frac{1}{1+rd} = \sum_{i \ne r} \frac{1}{1+id} = \frac{b}{c},$$

where c is not divisible by p. Hence, $caps_n - c = abp$. If s_n is an integer, this gives a contradiction since p does not divide c.

Next suppose that there exists one more integer m such that $u_m \equiv 0 \pmod{p}$ and $r < m \le n$. Then $(m-r)d \equiv 0 \pmod{p}$, and hence

$m - r \equiv 0 \pmod{p}$. If $m - r = tp$ and $t \geq 2$, then $m \geq r + 2p > n$. Hence,

$$t = 1, \quad m = r + p, \quad u_m = u_r + pd = (a + d)p,$$

if $u_r = ap$. This also shows that no other u_i, $1 \leq i \leq n$, is divisible by p. Thus,

$$s_n - \frac{1}{ap} - \frac{1}{(a+d)p} = \sum_{i \neq r, i \neq m} \frac{1}{1 + id} = \frac{b_1}{c_1},$$

where c_1 is not divisible by p. Hence,

$$c_1 a(a + d)ps_n - c_1(a + d + a) = a(a + d)b_1 p.$$

This shows that if s_n is an integer, then $p|(2a + d)$. Since $m \leq n \leq 2p$, we get

$$(a + d)p = 1 + md \leq 1 + 2pd, \quad (a - d)p \leq 1,$$

and hence, $a \leq d$. Thus, $2a + d \leq 3d < n/2 \leq p$. This contradicts the fact that $p|(2a + d)$.

We have thus arrived at the desired contradiction when $n \geq 64$. So let $n \leq 63$. As before, let $2^k \leq n < 2^{k+1}$. Then $k \leq 5$, and hence, $1 \leq s_n < (k + 1)/d \leq 6/d$ from (*). So $d < 6$.

Suppose that $q = u_r = 1 + rd$ is a prime for some $r \leq n$ and $r + q > n$. If $1 + md \equiv 0 \pmod{q}$ with $m > r$, then $md - rd \equiv 0 \pmod{q}$, which implies that

$$m - r \equiv 0 \pmod{q}, \quad m - r \geq q, \quad m \geq r + q > n.$$

So, if q is chosen like this, then no u_i, with $i \neq r$, is divisible by q. As before, we see that $s_n - 1/q = b/c$, where $\gcd(q, c) = 1$, and this gives a contradiction.

We have to verify that q can be chosen similarly in each of the special cases to be considered. The case $d = 1$ has been considered in an earlier problem.

Let $d = 2$: Then, $u_i = 2i + 1$ and $s_2 < 1$. If $3 \leq n \leq 9$, we can choose $q = u_3 = 7$. If $10 \leq n \leq 27$, let $q = u_9 = 19$; and if $28 \leq n \leq 63$, let $q = u_{26} = 53$.

Let $d = 3$: Then, $u_i = 3i+1$ and $s_2 < 1$. If $2 \leq n \leq 8$, let $q = u_2 = 7$. If $9 \leq n \leq 24$, let $q = u_6 = 19$; and if $25 \leq n \leq 63$, let $q = u_{20} = 61$.

Let $d = 4$: Then, $u_i = 4i + 1$ and $s_3 < 1$. If $4 \leq n \leq 20$, let $q = u_4 = 17$. If $21 \leq n \leq 63$, let $q = u_{15} = 61$.

Let $d = 5$: Then, $u_i = 5i + 1$ and $s_2 < 1$. If $2 \leq n \leq 12$, let $q = u_2 = 11$. If $13 \leq n \leq 63$, let $q = u_{12} = 61$. This completes the proof.

23. Prove that n is expressible as a sum of consecutive positive integers if and only if n is not a power of 2.

Solution A positive integer n is a power of 2 if and only if it has no odd divisor.

Suppose that $n = m + (m+1) + \cdots + (m+k-1)$. Then, $2n = k(2m+k-1)$. Note that if k is even, then $2m+k-1$ is odd. Hence, in any case, $2n$ has an odd divisor. On the other hand, if $n = Np$, where p is odd, then the sum of p consecutive integers beginning with $m = N - (p-1)/2$ is $pN = n$. If $m \leq 0$, we can remove the terms from m to $|m|$ without altering the sum. We see that n is expressible as the sum of consecutive positive integers if n has an odd divisor.

24. Let $k \geq 3$ be a fixed integer. Find all finite sequences $a_1, a_2, \ldots, a_k$ of positive integers such that the sum of any three terms in it is divisible by each of these three terms.

Solution Consider a finite sequence $a_1, a_2, \ldots, a_k$ of positive integers satisfying the condition. By permuting the terms we get another sequence with the same property. So we may suppose that $a_1 \leq a_2 \leq \ldots \leq a_k$.

If $a_1 = a_k$, then all the terms are equal. So let $a_1 < a_k$ and $2 \leq i \leq k-1$. Then, a_1, a_i, a_k are three terms in the sequence whose sum is divisible by a_1, a_i and a_k. So there exist positive integers m, n such that

$$a_1 + a_i + a_k = ma_i = na_k. \qquad (*)$$

Since $a_k < a_1 + a_i + a_k < 3a_k$, we get $a_k < na_k < 3a_k$, and hence $1 < n < 3$. It follows that $n = 2$, and $a_1 + a_i = a_k$. This holds whenever $2 \leq i \leq k-1$. Thus, we see that a_i has the same value for $2 \leq i \leq k-1$ and that $a_k = a_1 + a_2$.

If $a_1 = a_2$, then $a_1, \ldots a_{k-1}$ are all equal, and $a_k = 2a_1$; the sequence becomes $a_1, \ldots, a_1, 2a_1$.

Let $a_1 < a_2$. Then (*) with $i = 2$ shows that

$$ma_2 = a_1 + a_2 + a_k = 2a_1 + 2a_2.$$

Thus, $2a_2 < ma_2 < 4a_2$, which forces $m = 3$ and $a_2 = 2a_1$. If $k > 3$, then $2 < k - 1 < k$, and so $a_2 + a_{k-1} + a_k$ must be divisible by a_k. But

$$a_2 + a_{k-1} = a_2 + a_2 = 4a_1 \quad \text{and} \quad a_k = a_1 + a_2 = 3a_1.$$

So $a_2 + a_{k-1} + a_k = 7a_1$. Since $7a_1$ is not divisible by $3a_1 = a_k$, we cannot have $k > 3$ in the present case, $a_1 < a_2$.

Thus, the finite sequence $a_1, \ldots, a_k$ is in any of the following three forms:

(i) $a_1, a_1, \ldots, a_1$ with any $k \geq 3$.

(ii) $a_1, \ldots, a_1, 2a_1$, or a permutation of this, with any $k \geq 3$.

(iii) $k = 3$ and $a_1, 2a_1, 3a_1$.

25. Prove that there is no non-constant polynomial $f(x)$ with integer coefficients such that $f(x)$ is prime for every positive integer x.

Solution Let $f(x)$ be a non-constant polynomial with integer coefficients, a be any positive integer, and suppose that $f(a) = p$ is prime. We have $f(x) = (x-a)g(x) + f(a)$, where $g(x)$ is a polynomial with integer coefficients. Thus,

$$f(a + tp) = tp\, g(a + tp) + p, \quad \text{for any integer } t.$$

Now, $g(x)$ is an integer for all integral values of x. So $g(a + tp)$ is an integer for all integers t. This shows that p divides $f(a + tp)$ for all integers t. It follows that if $f(x)$ is prime for every positive integer x, then $f(a + tp) = p$ for all positive integers t. This cannot be true since a non-constant polynomial cannot have an infinite number of roots.

26. Find all triples (a, b, c) of non-zero integers satisfying the following conditions:

$$a \equiv b \pmod{|c|}, \quad b \equiv c \pmod{|a|}, \quad c \equiv a \pmod{|b|},$$
$$\gcd(a, b, c) = 1, \quad |a| \leq |b| \leq c.$$

Solution Suppose that (a, b, c) satisfies the conditions. Then there exists integers x, y and z such that

$$a - b = cx, \quad b - c = ay, \quad c - a = bz. \tag{*}$$

This shows that a common divisor of any two of a, b, c will also divide the third. Since $\gcd(a, b, c) = 1$, we see that a, b, c are pair-wise relatively prime to each other. Further,

$$c|x| = |a - b| \le |a| + |b| \le 2c,$$

which implies that $|x| = 0$, 1 or 2.
(i) Let $x = \pm 2$. First observe that $|x| = 2$ only when there is equality in both places. Thus, $|a - b| = 2c \ne 0$, and $|a| = |b| = c$; so $b = -a$. Since $\gcd(a, b) = 1$ and $c = |a|$, we see that the triple is

$$(1, -1, 1) \quad \text{or} \quad (-1, 1, 1).$$

(ii) Let $x = 1$. Then, $a - b = c$ from (*). Since $a \le |a| \le |b| \le c$, we get

$$b = a - c \le 0, \quad a = c + b = c - |b| \ge 0.$$

Thus, $b \le -1$, $a \ge 1$, and from (*),

$$a + ay = (b + c) + (b - c) = 2b.$$

Hence $a|2b$. Since $\gcd(a, b) = 1$, we get $a|2$. Thus, $a = 1$ or 2, and $c = a - b \ge a + 1$. If $a = 2$, then b and c are odd as $\gcd(a, b) = \gcd(a, c) = 1$. Hence, the triple (a, b, c), in the present case, is

$$(1, 1 - n, n) \quad \text{or} \quad (2, 1 - 2n, 2n + 1), \quad \text{where } n \text{ is an integer} \ge 2.$$

(iii) Next let $x = -1$. Then $a - b = -c$ from (*). Since $b \le |b| \le c$, we get

$$-a = c - b \ge 0, \quad b = c + a = c - |a| \ge 0.$$

Thus, $a \le -1$, $b \ge 1$, and from (*),

$$b - bz = (c + a) - (c - a) = 2a.$$

Hence $b|2a$. Since $\gcd(a, b) = 1$, we get $b|2$. Thus, $b = 1$ or 2, as before. Further,

$$c - b = -a \le b \le c, \quad b \le c \le 2b.$$

Since $c - b = -a \neq 0$, we get $b < c \leq 2b$. It follows that if $b = 1$, then $c = 2$, $a = -1$, and if $b = 2$, then $c = 3$ as $\gcd(b, c) = 1$. So the triple is

$$(-1, 1, 2) \quad \text{or} \quad (-1, 2, 3).$$

(iv) Finally, let $x = 0$. Then $a = b$. Since $\gcd(a, b) = 1$, this implies that $|a| = |b| = 1$, and c can be any positive integer. In this case, the triple is

$$(1, 1, n) \quad \text{or} \quad (-1, -1, n), \quad \text{where } n \text{ is any positive integer.}$$

27. Find the smallest positive integer n such that $\phi(x) = n$ has no solution; exactly two solutions; exactly three solutions; exactly four solutions.

Solution We have $\phi(1) = \phi(2) = 1$, $\phi(3) = \phi(4) = \phi(6) = 2$, and $\phi(N)$ is even if $N > 2$. Thus, we see that the equation $\phi(x) = n$ has no solution if $n = 3$, and 3 is the least such integer.

The least value of n for which the equation $\phi(x) = n$ has exactly two solutions is clearly $n = 1$. Moreover, the equation $\phi(x) = 2$ has at least three solutions, namely $x = 3, 4$ and 6. Since $\phi(5) = \phi(8) = \phi(10) = \phi(12) = 4$, the equation $\phi(x) = 4$ has at least four solutions, namely $x = 5, 8, 10$ and 12. We shall next see that the equations $\phi(x) = 2$ and $\phi(x) = 4$ have no more solutions.

Suppose that m is odd. Then $\phi(m) = 1$ if and only if $m = 1$. Let $m > 1$, p be a prime, and $m = p^a m_1$, where $a \geq 1$ and m_1 is not divisible by p. Then, p^a is relatively prime to m_1, and hence $\phi(m) = \phi(p^a)\phi(m_1) = p^{a-1}(p-1)\phi(m_1)$. Thus, $p - 1$ is an even divisor of $\phi(m)$. Moreover, if $\phi(m)$ is not divisible by p, then $a = 1$. Hence, if $\phi(m) = 2$, then $p - 1 = 2$, $p = 3$, $a = 1$ and $m_1 = 1$. If $\phi(m) = 4$, then $p - 1 = 2$ or 4. The case $p - 1 = 2$ yields contradicting results, namely, $p = 3$ and $m_1 = 3$. Hence, $p - 1 = 4$, $p = 5$ and $m_1 = 1$. Thus, if m is odd, then

$$\phi(m) = 2 \Longrightarrow m = 3, \quad \phi(m) = 4 \Longrightarrow m = 5.$$

If N is even, then $N = 2^r m$, where $r \geq 1$, m is odd, and $\phi(N) = 2^{r-1}\phi(m)$. This is divisible by 2^{r-1}.

Consider the relation $\phi(N) = 2$. If N is odd, then $N = 3$. If $N = 2^r m$, where m is odd, then 2 is divisible by 2^{r-1}, and so $r - 1 \leq 1$. If $r = 1$, then $\phi(m) = 2$, $m = 3$, and $N = 6$. If $r = 2$, then $\phi(m) = 1$, $m = 1$, and $N = 4$. The equation $\phi(x) = 2$ has no more solutions.

Next consider the relation $\phi(N) = 4$. If N is odd, then $N = 5$. If $N = 2^r m$, then 4 is divisible by 2^{r-1}, and so $r - 1 \leq 2$. If $r = 1$, then $\phi(m) = 4$, $m = 5$, and $N = 10$. If $r = 2$, then $\phi(m) = 2$, $m = 3$, and $N = 12$. If $r = 3$, then $\phi(m) = 1$, $m = 1$, and $N = 8$. The equation $\phi(x) = 4$ has no more solutions.

Thus, the least values of n for which the equation $\phi(x) = n$ has zero, two, three and four solutions are $n = 3$, $n = 1$, $n = 2$ and $n = 4$, respectively.

28. Find all solutions x of $\phi(x) = 24$.

Solution Let $\phi(N) = 24$. If p is prime and $p^r|N$, then $p^{r-1}(p-1)$ divides $\phi(N) = 24$. Hence, $p|N \Rightarrow p - 1 = 1, 2, 3, 4, 6, 8, 12$ or 24, that is, $p = 2, 3, 5, 7$ or 13 since p is prime. Let $N = 2^a\, 3^b\, 5^c\, 7^d\, 13^e$. Further, if $p^r|N$, then $p^{r-1}|24$. It follows that $0 \leq a \leq 4$, $0 \leq b \leq 2$, $0 \leq c, d, e \leq 1$.

Let $e = 1$. Then $N = 13N_1$, $N_1 = 2^a\, 3^b\, 5^c\, 7^d$, $\phi(N_1) = 2$, and hence, $N_1 = 3, 4$ or 6, $N = 39, 52$ or 78.

Let $e = 0$ and $d = 1$. Then, $N = 7N_2$, $N_2 = 2^a 3^b 5^c$, $\phi(N_2) = 4$, and hence, $N_2 = 5, 8, 10$ or 12, $N = 35, 56, 70$ or 84.

Let $e = d = 0$. Then, $N = 2^a 3^b n$, where $n = 5$ or 1. So $24 = \phi(2^a)\phi(3^b)\phi(n)$. Now, $\phi(2^a) = 1$ or 2^{a-1}, and $\phi(3^b) = 1$ or $3^{b-1}2$. Since $3|24$ and $\phi(n) = 4$ or 1, we see that $3|3^{b-1}$, and hence $b = 2$. It follows that if $n = 5$, then $\phi(2^a) = 1$, and if $n = 1$, then $\phi(2^a) = 4$. Thus, $N = 2^a 3^2 n = 1 \cdot 9 \cdot 5, 2 \cdot 9 \cdot 5$ or $8 \cdot 9$, that is, $N = 45, 90$ or 72.

Hence, there are 10 solutions, namely,

$$N = 39, 52, 78, 35, 56, 70, 84, 45, 90, 72.$$

29. Suppose that $I(x)$ denotes the integer nearest to x, where we choose the larger one if there are two such integers. Prove that for any positive integer n, we have

$$\sum_{k=1}^{\infty} I(n/2^k) = n.$$

Solution Let s_n denote the sum. We have to prove that $s_n = n$ for all positive integers n.

If $m - \frac{1}{2} \leq x < m + \frac{1}{2}$, where m is an integer, then $I(x) = m$. It follows that $I(x) = \left[x + \frac{1}{2}\right]$. Thus, $I(x) = 0$ for $0 < x < \frac{1}{2}$. Hence, if

$n < 2^r$, then $I\left(\frac{n}{2^k}\right) = 0$ for $k \geq r+1$, and $s_n = \sum_{k=1}^{r} I\left(\frac{n}{2^k}\right)$. Thus, $s_1 = I\left(\frac{1}{2}\right) = \left[\frac{1}{2} + \frac{1}{2}\right] = 1$, and

$$s_2 = I\left(\frac{2}{2}\right) + I\left(\frac{2}{4}\right) = \left[\frac{2}{2} + \frac{1}{2}\right] + \left[\frac{2}{4} + \frac{1}{2}\right] = 1 + 1 = 2.$$

Assume that $s_m = m$ for some m, and let $m < 2^r$. Then, $2m < 2m+1 < 2^{r+1}$. Hence,

$$s_{2m} = \sum_{k=1}^{r+1} I\left(\frac{2m}{2^k}\right), \quad s_{2m+1} = \sum_{k=1}^{r+1} I\left(\frac{2m+1}{2^k}\right).$$

Suppose that $I\left(\frac{2m}{2^k}\right) = \left[\frac{2m}{2^k} + \frac{1}{2}\right] = N$. Then,

$$N \leq \frac{2m}{2^k} + \frac{1}{2} < N+1, \quad N2^k \leq 2m + 2^{k-1} < (N+1)2^k.$$

If $k \geq 2$, then $2m+1+2^{k-1}$ is odd, and so $2m+1+2^{k-1} < (N+1)2^k$. This shows that if $k \geq 2$, then

$$N < \frac{2m+1}{2^k} + \frac{1}{2} < N+1,$$

and hence,

$$I\left(\frac{2m+1}{2^k}\right) = \left[\frac{2m+1}{2^k} + \frac{1}{2}\right] = N = I\left(\frac{2m}{2^k}\right).$$

Thus,

$$\sum_{k=2}^{r+1} I\left(\frac{2m+1}{2^k}\right) = \sum_{k=2}^{r+1} I\left(\frac{2m}{2^k}\right) = \sum_{k=2}^{r+1} I\left(\frac{m}{2^{k-1}}\right) = s_m.$$

Since $I\left(\frac{2m}{2}\right) = I(m) = m$ and $I\left(\frac{2m+1}{2}\right) = I\left(m + \frac{1}{2}\right) = m+1$, we see that $s_{2m} = m + s_m = 2m$ and $s_{2m+1} = m + 1 + s_m = 2m+1$, by our assumption. It follows by induction that $s_n = n$ for all n.

30. Let $f(x) = x - [x] - \frac{1}{2}$. Prove that

$$\left|\sum_{n=1}^{m} f\left(2^n x + \frac{1}{2}\right)\right| \leq 1 \quad \text{for all } m \geq 1 \text{ and all real } x.$$

Solution Let $k = [y]$. If $k \leq y < k + 1/2$, then $[y + 1/2] = k$ and $[2y] = 2k$. If $k + 1/2 \leq y < k + 1$, then $[y + 1/2] = k + 1$ and $[2y] = 2k + 1$. In either case, we have $[y] + [y + 1/2] = [2y]$, that is, $[y + 1/2] = [2y] - [y]$.

This with $y = 2^n x$ shows that $\left[2^n x + \frac{1}{2}\right] = [2^{n+1}x] - [2^n x]$ for all n. Since $f\left(2^n x + \frac{1}{2}\right) = \left(2^n x + \frac{1}{2}\right) - \left[2^n x + \frac{1}{2}\right] - \frac{1}{2}$, we have

$$\sum_{n=1}^{m} f\left(2^n x + \tfrac{1}{2}\right) = \sum_{n=1}^{m} \left\{2^n x - \left[2^n x + \tfrac{1}{2}\right]\right\}$$

$$= x(2^{m+1} - 2) - \sum_{n=1}^{m} \left[2^n x + \tfrac{1}{2}\right].$$

The last sum is equal to $\sum_{n=1}^{m} \left([2^{n+1}x] - [2^n x]\right) = [2^{m+1}x] - [2x]$. Hence,

$$\sum_{n=1}^{m} f\left(2^n x + \tfrac{1}{2}\right) = \left\{2^{m+1}x - [2^{m+1}x]\right\} - \left\{2x - [2x]\right\} = u - v,$$

where u and v are numbers such that $0 \leq u, v < 1$. This proves the result.

Set VIII

1. If n is a positive integer, prove that

$$\left[\sqrt{n} + \sqrt{n+1}\right] = \left[\sqrt{4n+1}\right] = \left[\sqrt{4n+2}\right] = \left[\sqrt{4n+3}\right].$$

Determine all integers r such that

$$\left[\sqrt{n} + \sqrt{n+1}\right] = \left[\sqrt{4n+r}\right], \quad \text{for all } n \geq 1.$$

Solution Let $m = [\sqrt{4n+1}]$ and $k = [\sqrt{4n+3}]$. Then $m \leq k$, and

$$m^2 \leq 4n + 1 < (m+1)^2, \quad k^2 \leq 4n + 3 < (k+1)^2.$$

Note that $(m+1)^2 > 4n+1$ implies that $(m+1)^2 \geq 4n+4$ since the square of an integer cannot be congruent to 2 or 3 modulo 4. It follows that $(m+1)^2 > k^2$, that is, $m \geq k$. Thus $m = k$.

Again, $4n^2 < 4n(n+1) < (2n+1)^2$, and so $2n < 2\sqrt{n(n+1)} < 2n+1$. Adding $2n+1$ we get,

$$4n+1 < n+(n+1)+2\sqrt{n(n+1)} < 4n+2 < 4n+3,$$

$$m^2 < \left(\sqrt{n}+\sqrt{n+1}\right)^2 < (k+1)^2 = (m+1)^2,$$

Thus $m = \left[\sqrt{n}+\sqrt{n+1}\right]$ which completes the first part.

Let $n = k(k+1)$ for some k. Then, $4n+1 = 4(k^2+k)+1 = (2k+1)^2$, while $4n < (2k+1)^2$. Hence, if $r \leq 0$, then

$$\left[\sqrt{4n+r}\right] \leq \left[\sqrt{4n}\right] < 2k+1 = \left[\sqrt{4n+1}\right] = \left[\sqrt{n}+\sqrt{n+1}\right]$$

for this n. So we cannot have $r \leq 0$. If $n = k^2 - 1$, then $\sqrt{4n+4} = \sqrt{4k^2}$, whereas $\sqrt{4n+3} = \sqrt{4k^2-1} < 2k$; hence, if $r \geq 4$, then

$$\left[\sqrt{4n+r}\right] \geq \left[\sqrt{4n+4}\right] = 2k > \left[\sqrt{4n+3}\right] = \left[\sqrt{n}+\sqrt{n+1}\right].$$

So we cannot have $r \geq 4$. Thus, the possible values of r are 1, 2 and 3.

2. If n is a positive integer, prove that

$$\left[\frac{8n+13}{25}\right] - \left[\frac{n-12-[(n-17)/25]}{3}\right]$$

is independent of n.

Solution Let

$$k = \left[\frac{n-17}{25}\right], \quad x = \frac{8n+13}{25}, \quad y = \frac{n-12-k}{3}.$$

Then, $25k \leq n-17 < 25(k+1)$, and $n-12-k = 3y$. So,

$$24k+5 \leq 3y < 24k+30, \quad 8k+1 < y < 8k+10.$$

Hence, $[y] = 8k+r$, where $1 \leq r \leq 9$, so that $8k+r \leq y < 8k+r+1$. Now, $n = 3y+k+12$. Since $24k+3r \leq 3y < 24k+3r+3$, we get

$$25k+3r+12 \leq n \leq 25k+3r+14,$$

$$200k+24r+109 \leq 8n+13 \leq 200k+24r+125.$$

But $8n+13 = 25x$. It follows that

$$8k+r+4+\frac{9-r}{25} \leq x \leq 8k+r+5-\frac{r}{25}.$$

Since $1 \leq r \leq 9$, this shows that $8k+r+4 \leq x < 8k+r+5$. Hence, $[x] = 8k+r+4 = [y]+4$, and the given expression is equal to $[x]-[y] = 4$, independent of n.

3. Prove that, for $x > 0$, $\sum_{n \le x} \left[\sqrt{\frac{x}{n}}\right] = \sum_{n \le \sqrt{x}} \left[\frac{x}{n^2}\right]$.

Solution Let $k = \left[\sqrt{x/n}\right]$, where $n \le x$. Then $k^2 \le x/n < (k+1)^2$, and so $n \le x/k^2$ and $n > x/(k+1)^2$. Note that there are $[y]$ integers n with $1 \le n \le y$. Thus, if $a_k = \left[x/k^2\right]$, then there are a_k integers n with $1 \le n \le x/k^2$. It follows that $[x/n] = k$ for $a_k - a_{k+1}$ values of n. In other words, the sum on the left-hand side contains exactly $a_k - a_{k+1}$ terms, each of which is equal to k. Let $m = [\sqrt{x}\,]$ so that $m^2 \le x < (m+1)^2$. Since $a_k = 0$ when $x < k^2$, that is, when $k > m$, we see that the value of the sum on the left-hand side is

$$\sum_{k \le \sqrt{x}} k(a_k - a_{k+1}) = \sum_{k=1}^{m} \{ka_k - (k+1)a_{k+1} + a_{k+1}\}$$

$$= a_1 - (m+1)a_{m+1} + \sum_{k=1}^{m} a_{k+1}.$$

Since $a_1 = x$ and $a_{m+1} = \left[x/(m+1)^2\right] = 0$, we see that the last sum is the same as the sum on the right-hand side.

4. Prove that $\sum_{k=1}^{n} (2k-1)\left[\frac{n}{k}\right] = \sum_{k=1}^{n} \left[\frac{n}{k}\right]^2$ for all positive integers n.

Solution Let $r = [n/k]$ for $k \le n$. Then $r \le n/k < (r+1)$, so that $k \le n/r$ and $k > n/(r+1)$. Since there are $[x]$ integers k with $1 \le k \le x$, we see that if $a_r = [n/r]$, then the sum on the right-hand side contains exactly $a_r - a_{r+1}$ terms each of which is equal to r^2. Since $a_r = 0$ when $n < r$, we see that the value of this sum is

$$\sum_{r=1}^{n} r^2(a_r - a_{r+1}) = \sum_{r=1}^{n} \{r^2 a_r - (r+1)^2 a_{r+1} + (2r+1)a_{r+1}\}$$

$$= a_1 - (n+1)^2 a_{n+1} + \sum_{r=1}^{n} (2r+1)a_{r+1}.$$

Since $a_1 = n$ and $a_{n+1} = 0$, we see that the last sum is the same as the sum on the left-hand side.

5. Prove that (i) $\sum_{k=1}^{n} \left[\frac{k}{2}\right] = \left[\frac{n^2}{4}\right]$ and (ii) $\sum_{k=1}^{n} \left[\frac{k}{3}\right] = \left[\frac{n(n-1)}{6}\right]$, for all positive integers n.

Solution Let us denote the sum by s_n.

(i) If $n = 2m - 1$, then

$$s_n = \left[\frac{1}{2}\right] + \sum_{r=1}^{m-1}\left\{\left[\frac{2r}{2}\right] + \left[\frac{2r+1}{2}\right]\right\} = \sum_{r=1}^{m-1}(r+r)$$

$$= m(m-1) = \frac{n+1}{2}\frac{n-1}{2} = \frac{n^2}{4} - \frac{1}{4}.$$

If $n = 2m$, then

$$s_n = s_{2m-1} + \left[\frac{2m}{2}\right] = m(m-1) + m = m^2 = \frac{n^2}{4}.$$

Hence, in either case, $\left[\frac{n^2}{4}\right] = s_n$.

(ii) As before,

$$\sum_{r=1}^{m-1}\left\{\left[\frac{3r}{3}\right] + \left[\frac{3r+1}{3}\right] + \left[\frac{3r+2}{3}\right]\right\} = \sum_{r=1}^{m-1}(3r).$$

Hence, if $n = 3m - 1$, then

$$s_n = \frac{3m(m-1)}{2} = \frac{(n+1)(n-2)}{6} = \frac{n(n-1)}{6} - \frac{2}{6}.$$

If $n = 3m$, then

$$s_n = s_{3m-1} + m = \frac{3m(m-1)}{2} + m = \frac{m(3m-1)}{2} = \frac{n(n-1)}{6},$$

and if $n = 3m + 1$, then

$$s_n = s_{3m} + m = \frac{m(3m-1)}{2} + m = \frac{m(3m+1)}{2} = \frac{(n-1)n}{6}.$$

Hence, in all cases, $[n(n-1)/6] = s_n$. This completes the proof.

6. If $a = 1, 2, \ldots, 7$, prove that there exists an integer b (depending on a) such that

$$\sum_{k=1}^{n}\left[\frac{k}{a}\right] = \left[\frac{(2n+b)^2}{8a}\right] \quad \text{for all positive integers } n.$$

Solution As in the previous problem, we shall denote the sum by s_n.

Since $\displaystyle\sum_{i=0}^{a-1}\left[\frac{ar+i}{a}\right] = ar$, we get

$$s_{am-1} = \sum_{r=1}^{m-1}\sum_{i=0}^{a-1}\left[\frac{ar+i}{a}\right] = \sum_{r=1}^{m-1}(ar) = \frac{am(m-1)}{2}.$$

Let $n = am + k$, where $0 \le k \le a-1$. Then,

$$s_n = s_{am-1} + \sum_{i=0}^{k}\left[\frac{am+i}{a}\right] = \frac{am(m-1)}{2} + (k+1)m.$$

Since $am = n - k$, we get

$$\begin{aligned}2as_n &= am(am-a) + 2a(k+1)m = (n-k)(n-k-a+2k+2),\\ &= n^2 + n(2-a) - (k^2+2k-ak) = (n^2+bn) - (k^2+kb),\end{aligned}$$

where $b = 2-a$. Hence, $8as_n = (2n+b)^2 - (2k+b)^2$. Now, $2k+b > -a$. Since $k+1 \le a$, we also have $2k+b = 2k+2-a \le a$. Hence, $(2k+b)^2 \le a^2$, and

$$8as_n \le (2n+b)^2 = 8as_n + (2k+b)^2 \le 8as_n + a^2,$$

$$s_n \le \frac{(2n+b)^2}{8a} \le s_n + \frac{a}{8} < s_n + 1 \quad \text{for } a \le 7.$$

This shows that $\left[\dfrac{(2n+b)^2}{8a}\right] = s_n$.

7. Show that $\left[(1+\sqrt{3})^{2n}\right] + 1$ and $\left[(1+\sqrt{3})^{2n+1}\right]$ are both divisible by 2^{n+1}. Is this the highest power of 2 dividing both the numbers?

Solution Let $u = 1+\sqrt{3}$ and $v = 1-\sqrt{3}$. Using binomial theorem,

$$u^n + v^n = 2\left\{1 + \binom{n}{2}\cdot 3 + \binom{n}{4}\cdot 3^2 + \cdots\right\} = \text{an integer}.$$

We have $1 < \sqrt{3} < 2$, and so $0 < -v < 1$; hence, $0 < (-1)^n v^n < 1$ for all n. Since $u^n = (u^n+v^n) - v^n$ and $u^n + v^n$ is an integer, we see that $[u^n]$ is equal to $u^n + v^n - 1$ if n is even, and to $u^n + v^n$ if n is odd. Thus,

$$[u^{2n}] = u^{2n} + v^{2n} - 1, \quad [u^{2n+1}] = u^{2n+1} + v^{2n+1}.$$

Now, $u^2 = 4 + 2\sqrt{3}$ and $v^2 = 4 - 2\sqrt{3}$. Hence,

$$u^{2n} + v^{2n} = 2^n \left\{ \left(2 + \sqrt{3}\right)^n + \left(2 - \sqrt{3}\right)^n \right\} = 2^{n+1} a_n,$$

where

$$a_n = 2^n + \binom{n}{2} 2^{n-2} \cdot 3 + \binom{n}{4} 2^{n-4} \cdot 3^2 + \cdots. \tag{1}$$

Since a_n is an integer, we first observe that $[u^{2n}] + 1 = u^{2n} + v^{2n}$ is divisible by 2^{n+1}. Further,

$$\begin{aligned} u^{2n+1} + v^{2n+1} &= (1 + \sqrt{3})u^{2n} + (1 - \sqrt{3})v^{2n} \\ &= u^{2n} + v^{2n} + \sqrt{3}\,(u^{2n} - v^{2n}), \end{aligned}$$

and

$$u^{2n} - v^{2n} = 2^n \left\{ \left(2 + \sqrt{3}\right)^n - \left(2 - \sqrt{3}\right)^n \right\} = 2^{n+1} \sqrt{3}\, b_n,$$

where

$$b_n = \binom{n}{1} 2^{n-1} + \binom{n}{3} 2^{n-3} \cdot 3 + \binom{n}{5} 2^{n-5} \cdot 3^2 + \cdots. \tag{2}$$

Since a_n and b_n are integers, this shows that $[u^{2n+1}] = u^{2n+1} + v^{2n+1} = 2^{n+1}(a_n + 3b_n)$ is divisible by 2^{n+1}. We have proved that both the numbers $[(1 + \sqrt{3})^{2n}] + 1$ and $[(1 + \sqrt{3})^{2n+1}]$ are divisible by 2^{n+1}, proving the first part.

For the next part, suppose that both these numbers are divisible by a more higher power of 2 than 2^{n+1}. Then a_n and $a_n + 3b_n$ are even, that is a_n and b_n are even. If n is even, then the last term in (1) is $\binom{n}{n} 2^0 3^{n/2}$ which is odd. Since the remaining terms in (1) are even, a_n is odd when n is even. On the other hand, if n is odd, then the last term in (2) is $\binom{n}{n} 2^0 3^{(n-1)/2}$ which is odd. Since the remaining terms in (2) are even, b_n is odd when n is odd. Thus, in either case, one of a_n, b_n is odd. This implies that 2^{n+1} is the highest power of 2 that divides both the numbers $[(1 + \sqrt{3})^{2n}] + 1$ and $[(1 + \sqrt{3})^{2n+1}]$.

8. If $2^k - 1$ is prime ($k > 1$), show that $n = 2^{k-1}(2^k - 1)$ is perfect. Show also that every even perfect number is of this form.

Solution A positive integer n is perfect if $\sigma(n) = 2n$, where $\sigma(n)$ is the sum of all positive divisors of n.

Let $p = 2^k - 1$ be prime, and $n = 2^{k-1}(2^k - 1) = p2^{k-1}$. The function $\sigma(n)$ is known to be multiplicative, that is, $\sigma(mn) = \sigma(m)\sigma(n)$ whenever m and n are coprime. Hence,

$$\sigma(n) = \sigma(p)\sigma(2^{k-1}) = (p+1)(2^k - 1) = 2^k(2^k - 1) = 2n.$$

This means that n is perfect.

For the converse, suppose that n is an even perfect number. We can express n as $n = 2^r m$ for some $r \geq 1$ and an odd m. Then,

$$2n = \sigma(n) = \sigma(2^r)\sigma(m) = (2^{r+1} - 1)\sigma(m).$$

Thus, $2^{r+1}m = (2^{r+1} - 1)\sigma(m)$. Since $2^{r+1} - 1$ is relatively prime to 2^{r+1}, this implies that $2^{r+1} - 1$ divides m. Let $m = (2^{r+1} - 1)m_1$. It follows that

$$\sigma(m) = 2^{r+1}m_1 = (2^{r+1} - 1)m_1 + m_1 = m + m_1.$$

But m and m_1 are two distinct divisors of m. If $m_1 \neq 1$, then $\sigma(m) \geq 1 + m_1 + m$. So $\sigma(m) = m + m_1$ implies that $m_1 = 1$. Thus, we see that m has only two divisors, namely, 1 and m. This means that m is prime. Further, $m = (2^{r+1} - 1)m_1 = (2^{r+1} - 1)$. This completes the proof.

9. Let n be an even perfect number. Show that $n \equiv 6 \pmod{10}$ or $n \equiv 28 \pmod{100}$. Show also that if $n \equiv 6 \pmod{100}$, then $n = 6$.

Solution If n is an even perfect number, then $n = 2^{k-1}(2^k - 1)$, where k is such that $2^k - 1$ is prime. If $k = 2$, then $n = 6$. If k is an even number greater than 2, then $2^k - 1$, being divisible by $2^2 - 1$, cannot be prime. So k is of the form $4m + 1$ or $4m + 3$. First let $k = 4m + 1$. Then $2^{k-1} = 2^{4m} \equiv 1 \pmod 5$, and hence, $n \equiv 1(2 - 1) \pmod 5$. This implies that $n \equiv 1$ or $6 \pmod{10}$. But n is even, and so $n \equiv 6 \pmod{10}$.

Next let $k = 4m + 3$. Then,

$$2^{k-3} = 2^{4m} = (1 + 15)^m \equiv 1 + 15m \pmod{25}.$$

Let $2^{k-3} = 1 + 15m + 25N$. Then,

$$2^{k-1} = 4 + 60m + 100N, \quad 2^k = 8 + 120m + 200N,$$

and hence,

$$\begin{aligned} n &= (4 + 60m + 100N)(7 + 120m + 200N) \\ &\equiv 28 + 480m + 420m. \end{aligned}$$

Thus, $n \equiv 28 \pmod{100}$ in this case.

For the second part, suppose that $n = 100a + 6$. But $n = 2^{k-1}(2^k - 1)$ for some integer $k > 1$. So $2^{k-2}(2^k - 1) = 50a + 3$ is odd. This implies that $k = 2$ and $n = 6$.

10. Let n be an even perfect number, and $n > 6$. Prove that $n \equiv 1 \pmod 9$ and $n \equiv 4 \pmod 6$.

Solution We have $n = 2^{k-1}(2^k - 1)$, where $k > 2$ and $2^k - 1$ is prime. If k has a divisor d with $1 < d < k$, then $2^d - 1$ is a proper divisor of $2^k - 1$. Hence, if $2^k - 1$ is prime, then k is prime. So k is of the form $6r + 1$ or $6r + 5$. If $k = 6r + 1$, then $2^{k-1} = 2^{6r} = 64^r \equiv 1 \pmod 9$, and hence, $n \equiv 1(2 - 1) \pmod 9$. If $k = 6r + 5$, then $2^{k-1} = 2^4 \cdot 2^{6r} = 16 \cdot 64^r \equiv 7 \pmod 9$, and hence, $n \equiv 7(14 - 1) \pmod 9$. In either case, $n \equiv 1 \pmod 9$.

If $n \equiv 1 \pmod 9$, then $n \equiv 1 \pmod 3$, and hence, $n = 6m + 1$ or $6m + 4$, for some integer m. Since n is even, we have $n = 6m + 4 \equiv 4 \pmod 6$.

11. Let n be an even perfect number, and $n > 6$. Prove that n can be expressed as a sum of consecutive odd cubes.

Solution Let s_m be the sum of cubes of the first m odd natural numbers, and t_m be the sum of cubes of the first m natural numbers. Then

$$s_m = t_{2m} - \left\{2^3 + 4^3 + \cdots + (2m)^3\right\} = t_{2m} - 8t_m.$$

Using the formula for the sum of cubes of the first n natural numbers, we have $t_r = \{r(r+1)/2\}^2$. Hence,

$$\begin{aligned} s_m &= \{m(2m+1)\}^2 - 2\{m(m+1)\}^2 \\ &= m^2\left\{4m^2 + 4m + 1 - 2m^2 - 4m - 2\right\} = m^2\left\{2m^2 - 1\right\}. \end{aligned}$$

If n is an even perfect number, then $n = 2^{k-1}(2^k - 1)$, where $2^k - 1$ is prime. If $n > 6$, then k is odd. Let $m = 2^{(k-1)/2}$. Then, $m^2 = 2^{k-1}$, $2m^2 - 1 = 2^k - 1$, and so $s_m = m^2(2m^2 - 1) = n$. Thus, n is equal to the sum of the cubes of the first m odd natural numbers, where $m = 2^{(k-1)/2}$.

12. Prove that no proper divisor of a perfect number can be perfect.

Solution If m is a proper divisor of n, then $n = mk$, where $1 < k < n$. If $d_1 = 1, d_2, \ldots, d_r = m$ are the various positive divisors of m, then $kd_1, kd_2, \ldots, kd_r$ are divisors of n, none of which is 1. It follows that

$$\sigma(n) \geq 1 + kd_1 + kd_2 + \cdots + kd_r = 1 + k\sigma(m).$$

If n is perfect, then $\sigma(n) = 2n = 2mk$, and therefore, $2mk \geq 1 + k\sigma(m) > k\sigma(m)$. Thus, if n is perfect, then $\sigma(m) < 2m$, and so m is not perfect.

13. Prove that a perfect square cannot be a perfect number.

Solution Let n be perfect. If n is even then $n = 2^{k-1}(2^k - 1)$, where $p = 2^k - 1$ is prime. In that case, n is divisible by p but not by p^2, and hence n cannot be a perfect square.

So let n be odd. Note that n has an equal number of divisors which are less than $\sqrt{n}$ and those which are greater than $\sqrt{n}$. If n is a perfect square, then $\sqrt{n}$ also is a divisor, and so n has an odd number of divisors. Since each divisor of n is odd, we see that $\sigma(n)$, the sum of divisors, is odd. Thus, n cannot a perfect number if it is odd and is a perfect square.

14. Let $H(n)$ be the harmonic mean of the positive divisors of n. Prove that $H(n)$ is an integer if n is perfect. Does there exist numbers n for which $H(n)$ is an integer, but n is not perfect?

Solution Let n be a perfect number, and $d_1 = 1, d_2, \ldots, d_k = n$ be the positive divisors of n so that $k = d(n)$ is the number of divisors of n. Then, $H(n)$ is given by

$$\frac{1}{H(n)} = \frac{1}{k}\left(\frac{1}{d_1} + \frac{1}{d_2} + \cdots + \frac{1}{d_k}\right) = \frac{1}{nd(n)}\left(\frac{n}{d_1} + \frac{n}{d_2} + \cdots + \frac{n}{d_k}\right).$$

But $n/d_1, n/d_2, \ldots, n/d_k$ are precisely the various divisors of n, written in a different order. It follows that

$$\frac{1}{H(n)} = \frac{1}{nd(n)}\sigma(n) = \frac{1}{nd(n)}2n = \frac{2}{d(n)}.$$

Thus, $H(n) = d(n)/2$. Note that the number of divisors of n is odd only when n is a perfect square. A perfect number cannot be a perfect square. Hence, $d(n)$ is even, and $H(n)$ is an integer.

The second part: We have proved above that $H(n) = nd(n)/\sigma(n)$, where $d(n)$ is the number of positive divisors of n. Thus, $H(n)$ is an integer if (i) $\sigma(n) = 2n$ and $d(n)$ is even, or (ii) $\sigma(n) = 3n$ and $d(n)$ is divisible by 3, or (iii) $\sigma(n) = 4n$ and $d(n)$ is divisible by 4, and so on. In case (i), n is perfect. If $n = 2^5 \cdot 3 \cdot 7$, we have $\sigma(n) = 63 \cdot 4 \cdot 8 = 3n$, and $d(n) = 6 \cdot 2 \cdot 2$. If $n = 2^5 \cdot 3^3 \cdot 5 \cdot 7$, we have $\sigma(n) = 63 \cdot 40 \cdot 6 \cdot 8 = 4n$, and $d(n) = 6 \cdot 4 \cdot 2 \cdot 2$. So the answer to the second question is 'Yes'.

15. (a) Prove that no power of a prime can be a perfect number.
    (b) If n is an odd perfect number, prove that n has at least three distinct prime divisors.

Solution (a) Let $n = p^a$, where p is a prime. Then

$$\sigma(n) = \frac{p^{a+1} - 1}{p - 1} < \frac{p^{a+1}}{p - 1} = n\frac{p}{p - 1} \leq 2n;$$

hence, $\sigma(n) < 2n$, and n is not perfect.

(b) If n has only one prime divisor, then n is not perfect by (a). Let n be odd, and $n = p^a q^b$, where p and q are distinct odd primes. Then,

$$\sigma(n) = \frac{p^{a+1} - 1}{p - 1} \cdot \frac{q^{b+1} - 1}{q - 1} < \frac{p^{a+1}}{p - 1} \cdot \frac{q^{b+1}}{q - 1} = n\frac{p}{p - 1} \cdot \frac{q}{q - 1}.$$

If $p < q$, then $p \geq 3$ and $q \geq 5$; hence,

$$\begin{aligned}\frac{p}{p-1} \cdot \frac{q}{q-1} &= \left(1 + \frac{1}{p-1}\right)\left(1 + \frac{1}{q-1}\right) \\ &\leq \left(1 + \frac{1}{2}\right)\left(1 + \frac{1}{4}\right) < 2.\end{aligned}$$

Thus, $\sigma(n) < 2n$, and n is not perfect. This shows that if n is odd and has less than three distinct prime divisors, then n is not perfect.

16. If n is an odd perfect number, prove that n can be expressed as $n = p_1^{k_1} p_2^{k_2} \cdots p_r^{k_r}$, where the p_i are distinct primes, $p_1 \equiv k_1 \equiv 1 \pmod 4$ and $k_2, \ldots, k_r$ are even.

Solution Let $n = p_1^{k_1} p_2^{k_2} \cdots p_r^{k_r}$, where the p_i are distinct odd primes. Since n is perfect, we have

$$2n = \sigma(n) = \sigma(p_1^{k_1})\sigma(p_2^{k_2}) \cdots \sigma(p_r^{k_r}).$$

Since n is odd, this shows that exactly one of the factors $\sigma(p_i^{k_i})$ is even and the remaining are odd. Further, the even factor is not divisible by 4.

If k_i is odd, then $p_i^{k_i}$ has $k_i + 1$ divisors $1, p_i, p_i^2, \ldots$, each of which is odd, and hence their sum is even, that is, $\sigma(p_i^{k_i})$ is even. It follows that k_i is odd for exactly one i, and the remaining k_i are even.

Let us rename the primes p_i, if necessary, so that k_1 is odd and $k_2, \ldots k_r$ are even. Thus, $\sigma(p_1^{k_1}) \equiv 2 \pmod 4$. Since p_1 is odd, we have $p_1 \equiv \pm 1 \pmod 4$. If $p_1 \equiv -1 \pmod 4$, then

$$\sigma(p_1^{k_1}) = (1 + p_1) + p_1^2(1 + p_1) + \cdots + p_1^{k_1 - 1}(1 + p_1)$$

is divisible by 4. This is not possible. Hence, $k_1 \equiv 1 \pmod 4$. This completes the proof.

In the following problems, u_n denotes the nth Fibonacci number, defined by $u_1 = u_2 = 1$, $u_{n+1} = u_n + u_{n-1}$ for $n > 2$.

17. Prove that

(a) $u_{m+n} = u_{m-1}u_n + u_m u_{n+1}$ for all $m \geq 2, n \geq 1$,

(b) $u_{n+2}^2 - u_n^2 = u_{2n+2}$.

Solution (a) For all $m \geq 2$, we have

$$u_{m-1}u_1 + u_m u_2 = u_{m-1} + u_m = u_{m+1},$$
$$u_{m-1}u_2 + u_m u_3 = u_{m-1} + 2u_m = u_{m+1} + u_m = u_{m+2}.$$

These show that the result holds when $n = 1$ and $n = 2$. Assume that it is true when $n = k$ and $n = k + 1$. Thus,

$$u_{m+k} = u_{m-1}u_k + u_m u_{k+1},$$
$$u_{m+k+1} = u_{m-1}u_{k+1} + u_m u_{k+2}.$$

On addition, we get

$$u_{m+k+2} = u_{m-1}u_{k+2} + u_m u_{k+3}.$$

So the result holds when $n = k + 2$. This proves (a) by induction.

(b) Applying (a) with m and n changed to $n + 1$ and $n + 1$, we get

$$u_{2n+2} = u_n u_{n+1} + u_{n+1}u_{n+2} = u_{n+1}(u_n + u_{n+2})$$
$$= (u_{n+2} - u_n)(u_n + u_{n+2}) = u_{n+2}^2 - u_n^2.$$

18. Prove that

 (a) $2^{n-1}u_n \equiv n \pmod 5$ for all $n \geq 1$,
 (b) $u_1^2 + u_2^2 + \cdots + u_n^2 = u_n u_{n+1}$ for $n \geq 1$.

Solution (a) Since $u_1 = 1 = u_2$, we see that the result holds when $n = 1$ and $n = 2$. If

$$2^{n-2}u_{n-1} \equiv n-1 \quad \text{and} \quad 2^{n-1}u_n \equiv n \pmod 5,$$

then

$$\begin{aligned} 2^n u_{n+1} &= 2 \cdot 2^{n-1}u_n + 4 \cdot 2^{n-2}u_{n-1} \\ &\equiv 2n + 4(n-1) \equiv n+1 \pmod 5. \end{aligned}$$

This proves (a) by induction.

(b) If $u_n u_{n+1}$ is denoted by x_n, then, for all $n \geq 2$,

$$x_n - x_{n-1} = u_n(u_{n+1} - u_{n-1}) = u_n^2.$$

Now, $u_1^2 = 1 = x_1$. Thus, the result holds when $n = 1$, and hence for all n.

19. Prove that, for $m, n \geq 1$, we have

 (a) $u_{5n+2} > 10^n$ for all $n \geq 1$,
 (b) u_{mn} is divisible by u_m,
 (c) u_{3n} is even, u_{4n} is divisible by 3, and u_{5n} is divisible by 5.

Solution (a) First observe that $u_7 = 13 > 10$. Next assume that $u_{5n+2} > 10^n$. Using Problem ??(a) with $m = 5n+2$ and $n = 5$, we get $u_{5n+7} = 5u_{5n+1} + 8u_{5n+2}$. But,

$$5u_{5n+1} > 4u_{5n+1} > 2(u_{5n+1} + u_{5n}) = 2u_{5n+2}.$$

It follows that $u_{5n+7} > 10u_{5n+2} > 10^{n+1}$. This proves the result by induction.

(b) This is trivial when $m = 1$ or $n = 1$. Let $m \geq 2$. Again, by Problem ??(a), we get

$$u_{mn+m} = u_{mn-1}u_m + u_{mn}u_{m+1}.$$

Hence, if u_{mn} is divisible by u_m, then $u_{m(n+1)}$ is also divisible by u_m. This gives (b) by induction.

(c) This follows from (b) since $u_3 = 2$, $u_4 = 3$ and $u_5 = 5$.

20. Prove that

(a) $u_m u_n - u_{m+1}u_{n-1} = (-1)^{n-1}u_{m-n+1}$ for $m \geq n \geq 2$,

(b) $u_n^2 - u_{n+1}u_{n-1} = (-1)^{n-1}$ for $n \geq 2$,

(c) if m, n and $m - n$ are positive integers, then a common divisor of u_m and u_n is also a divisor of u_{m-n}.

Solution (a) If $n = 2$, this becomes $u_m - u_{m+1} = -u_{m-1}$, which is obvious. So let $n \geq 3$. We have

$$\begin{aligned} u_m u_n - u_{m+1}u_{n-1} &= u_m(u_{n-1} + u_{n-2}) - (u_m + u_{m-1})u_{n-1} \\ &= -(u_{m-1}u_{n-1} - u_m u_{n-2}). \end{aligned}$$

By repeated application of this with m, n replaced by $m-1, n-1$, we get

$$u_m u_n - u_{m+1}u_{n-1} = (-1)^{n-2}(u_{m-n+2}u_2 - u_{m-n+3}u_1),$$

which proves the result since $u_2 = u_1 = 1$.

(b) This is a special case of (a) with $m = n$.

(c) Since $u_m u_{n+1} - u_{m+1}u_n = (-1)^n u_{m-n}$, by (a), we get the result.

21. Prove that, for all positive integers m, n,

(a) gcd $(u_m, u_n) = u_d$, where $d = \gcd(m, n)$,

(b) $u_m | u_n$ if and only if $m|n$.

Solution (a) Let $e = \gcd(u_m, u_n)$. If $m = m_1 d$ and $n = n_1 d$, then by Problem ??(b), we see that $u_m = u_{m_1 d}$ and $u_n = u_{n_1 d}$ are divisible by u_d. Consequently, $u_d | e$.

To prove that $e | u_d$, note that there exist positive integers a and b such that $d = am - bn$. The fact that $e | u_m$ and $u_m | u_{am}$ implies that e divides u_{am}, and hence, e divides u_{am-n}, by (c) of the previous problem. If $am - 2n > 0$, then one more application of this result shows that e divides u_{am-2n}. We can repeat the argument until we find that e divides $u_{am-bn} = u_d$. This finishes the proof.

(b) If $u_m | u_n$, then gcd $(u_m, u_n) = u_m$, that is, $u_d = u_m$, because of (a). This implies that $d = m$ and $m|n$. On the other hand, if $m|n$, let $n = mk$. Then, $u_n = u_{mk}$ is divisible by u_m by Problem ??(b).

22. Prove that there is no positive integer n for which $u_1 + u_2 + \cdots + u_{3n} = 16\,!$.

Solution We have $u_k = u_{k+2} - u_{k+1}$ for all k. Adding we get,

$$u_1 + u_2 + \cdots + u_N = u_{N+2} - u_2 = u_{N+2} - 1.$$

Hence, if $u_1 + u_2 + \cdots + u_{3n} = 16\,!$, then $u_{3n+2} - 1 = 16\,!$. Since $16\,! \equiv -1 \pmod{17}$, this implies that 17 divides u_{3n+2}. Since $u_9 = 34$ is divisible by 17, we see that if 17 divides u_m, then 17 divides (u_m, u_9). But $(u_m, u_9) = u_{(m,9)}$, and $(m, 9) = 9, 3$ or 1. Since 17 does not divide u_3 or u_1, we see that if $17 | u_m$, then $(m, 9) = 9$. It follows that 17 does not divide u_{3n+2}. This proves the result.

23. Prove that, for any positive integer n, we can find n consecutive composite Fibonacci numbers.

Solution Consider the n consecutive composite numbers $a_k = (n + 2)\,! + k$, $3 \le k \le n + 2$. Since a_k is divisible by k, we see, by Problem **??**(b), that u_{a_k} is divisible by u_k. Further, $2 < k < a_k$, and so $1 < u_k < u_{a_k}$. Hence, each u_{a_k} is composite.

24. Prove that any positive integer N can be expressed as a sum of distinct Fibonacci numbers, no two of which are consecutive.

Solution A 'sum' may consist of just one term. Let us first observe that $1 = u_2$, $2 = u_3$, $3 = u_4$, $4 = u_4 + u_2$, and so on. We shall first prove that every positive integer less than u_n is expressible as a sum of the distinct elements in the set $S = \{u_2, u_3, \ldots, u_{n-1}\}$. This holds when $n = 3$, 4 and 5. Assume that it is true for some n, and let $u_n < N < u_{n+1}$. Then,

$$0 < N - u_n < u_{n+1} - u_n = u_{n-1} < u_n.$$

Hence, $N - u_n$ is a sum of the distinct elements in S. This shows that $N = (N - u_n) + u_n$ is a sum of distinct elements in $\{u_2, \ldots, u_n\}$ and proves the assertion, by induction.

Such an expression may involve consecutive Fibonacci numbers. If u_k and u_{k-1} are two consecutive Fibonacci numbers included in the sum, and u_{k+1} is not included in it, then $u_k + u_{k-1}$ may be replaced by u_{k+1}. This process reduces the number of terms in the sum. It can be repeated until we get rid of the consecutive Fibonacci numbers in the sum.

25. Let $\alpha = \frac{1}{2}(1 + \sqrt{5})$ and $\beta = \frac{1}{2}(1 - \sqrt{5})$. Prove that, for all $n \ge 1$,

(a) $u_n = \dfrac{\alpha^n - \beta^n}{\sqrt{5}}$,

(b) u_n is the nearest integer to $\dfrac{\alpha^n}{\sqrt{5}}$.

Solution (a) Since $\alpha + \beta = 1$, we have $\alpha^2 - \beta^2 = \alpha - \beta = \sqrt{5}$. This shows that the result is true when $n = 1$ and $n = 2$. So let $n \geq 2$, and x be a real number such that $x^2 = x + 1$. Then $x^3 = x^2 + x = 2x + 1$. We shall prove that $x^n = u_n x + u_{n-1}$ for $n \geq 2$. This is verified for $n = 2, 3$. Assume that it is true when $n = k$ and $n = k + 1$. Then,

$$x^{k+2} = x^{k+1} + x^k = (u_{k+1}x + u_k) + (u_k x + u_{k-1}) = u_{k+2}x + u_{k+1}.$$

This proves our assertion by induction.

Since α and β satisfy the equation $x^2 = x + 1$, we get

$$\alpha^n = u_n\alpha + u_{n-1}, \quad \beta^n = u_n\beta + u_{n-1}.$$

Hence, $\alpha^n - \beta^n = u_n(\alpha - \beta) = u_n(\sqrt{5})$, as is to be proved.

(b) Since $2 < \sqrt{5} < 3$, we can see that $-1 < \beta < -\frac{1}{2}$, and hence, $|\beta^n| < 1$. Thus,

$$\left|\frac{\alpha^n}{\sqrt{5}} - u_n\right| = \left|\frac{\beta^n}{\sqrt{5}}\right| < \frac{1}{\sqrt{5}} < \frac{1}{2}, \quad u_n - \frac{1}{2} < \frac{\alpha^n}{\sqrt{5}} < u_n + \frac{1}{2}.$$

This proves (b).

26. Let p be a prime and $p > 5$. Prove that either $p|u_{p-1}$ or $p|u_{p+1}$.

Solution By the formula for u_n given in the previous problem, we have $\sqrt{5}\, u_p = \alpha^p - \beta^p$, where $\alpha = \frac{1}{2}(1+\sqrt{5})$ and $\beta = \frac{1}{2}(1-\sqrt{5})$. It follows that

$$u_p = \frac{1}{2^{p-1}}\left\{\binom{p}{1} + \binom{p}{3}5 + \binom{p}{5}5^2 + \cdots + \binom{p}{p}5^{(p-1)/2}\right\}.$$

For $1 \leq k \leq p - 1$, the binomial coefficients $\binom{p}{k}$ are divisible by p. Thus, we see that $2^{p-1}u_p \equiv 5^{(p-1)/2} \pmod{p}$, and so $4^{p-1}u_p^2 \equiv 5^{p-1} \pmod{p}$. But $4^{p-1} \equiv 1 \equiv 5^{p-1} \pmod{p}$. Thus, $u_p^2 \equiv 1 \pmod{p}$.

By Problem **??**(b), $u_p^2 - u_{p-1}u_{p+1} = 1$. This shows that $u_{p-1}u_{p+1} \equiv 0 \pmod{p}$. This proves the result.

27. Prove that the product $u_n u_{n+1} u_{n+2} u_{n+3}$ of any four consecutive Fibonacci numbers is equal to the area of a Pythagorean triangle.

Solution If $x = r^2 - s^2$, $y = 2rs$ and $z = r^2 + s^2$, then $x^2 + y^2 = z^2$; that is, (x, y, z) is a Pythagorean triple. If $r = u_{n+2}$ and $s = u_{n+1}$, then

$$r^2 - s^2 = (u_{n+2} - u_{n+1})(u_{n+2} + u_{n+1}) = u_n u_{n+3}.$$

Hence, the area of the Pythagorean triangle with sides x, y, z is

$$\frac{1}{2}xy = (r^2 - s^2)rs = u_n u_{n+3} u_{n+2} u_{n+1}.$$

28. If m is an even positive integer, prove that the sum of any $2m$ consecutive Fibonacci numbers is divisible by u_m. Also determine all odd integers m with this property. Deduce that the sum of any 24 consecutive Fibonacci numbers is divisible by 24.

Solution We have $u_{n+2} - u_{n+1} = u_n$. On addition, we get

$$s_n = u_1 + u_2 + \cdots + u_n = u_{n+2} - u_2.$$

Hence, the sum of $2m$ consecutive numbers $u_{n+1}, \ldots, u_{n+2m}$ is $s_{n+2m} - s_n = u_{n+2m+2} - u_{n+2}$. Using Problem **??**(a), we can see that this sum is

$$u_{n+1}u_{2m} + u_{n+2}u_{2m+1} - u_{n+2} = u_{n+1}u_{2m} + u_{n+2}(u_{2m+1} - 1).$$

Since u_{2m} is divisible by u_m, we see that this sum is divisible by u_m if and only if $u_{2m+1} - 1$ is divisible by u_m. Now, $u_{2m+1} = u_m u_m + u_{m+1}u_{m+1}$, and $u_{m+1}^2 - u_m u_{m+2} = (-1)^m$, by Problem **??**(b). Hence,

$$u_{2m+1} - 1 = u_m^2 + u_{m+1}^2 - 1 = u_m^2 + u_m u_{m+2} + (-1)^m - 1.$$

If m is even, then $(-1)^m - 1 = 0$ and so the sum is divisible by u_m. If m is odd, then $(-1)^m - 1 = -2$ and so u_m divides this sum if and only if it divides -2, that is, $u_m = 1$ or 2. Hence, the only odd values of m with this property are $m = 1, 3$.

For the last statement, apply the above result with $m = 12$, and note that $u_{12} = 144$ is divisible by 24.

29. Determine all integers n such that $u_n + 1$ is prime.

Solution Observe that $u_1 + 1 = u_2 + 1 = 2$, $u_3 + 1 = 3$, $u_4 + 1 = 4$, and $u_5 + 1 = 6$. Thus, $u_n + 1$ is prime when $n = 1$, 2 and 3, and is composite when $n = 4$ and 5. So suppose that $n > 5$.

Let $n = 2k$, where $k \geq 3$. By Problem ??(a), we have

$$u_{2k} = u_{k-1}u_k + u_k u_{k+1}. \qquad (*)$$

Now, using Problem ??(b), we have

$$u_k u_{k+1} = u_k^2 + u_k u_{k-1} = u_{k+1}u_{k-1} + (-1)^{k-1} + u_k u_{k-1},$$
$$u_{k-1}u_k = u_{k-1}^2 + u_{k-1}u_{k-2} = u_k u_{k-2} + (-1)^{k-2} + u_{k-1}u_{k-2}.$$

It follows from (*) that, if k is even, then

$$u_{2k} + 1 = u_{k-1}u_k + u_{k+1}u_{k-1} + u_k u_{k-1}$$

is divisible by u_{k-1}. If k is odd, then

$$u_{2k} + 1 = u_k u_{k-2} + u_{k-1}u_{k-2} + u_k u_{k+1} = u_{k-2}u_{k+1} + u_k u_{k+1}$$

is divisible by u_{k+1}. In either case, it is composite for all $k \geq 3$.

Next, consider the case $n = 2k + 1$, where $k \geq 3$. As before, we have $u_{2k+1} = u_k u_k + u_{k+1}u_{k+1}$. Using Problem ??(b), we can write this in two ways as

$$u_{2k+1} = u_{k+1}u_{k-1} + (-1)^{k-1} + u_{k+1}^2,$$
$$u_{2k+1} = u_k^2 + u_{k+2}u_k + (-1)^k.$$

If k is even, the first one shows that $u_{2k+1} + 1$ is divisible by u_{k+1}. If k is odd, then $u_{2k+1} + 1$ is divisible by u_k, by the second equation. In either case, it is composite.

Thus, we see that $u_n + 1$ is prime when $n = 1, 2$ and 3, and is composite for all $n \geq 4$.

30. Prove that, for $n \geq 2$,

$$u_n = \binom{n-1}{0} + \binom{n-2}{1} + \binom{n-3}{2} + \cdots,$$

where we use the convention that $\binom{m}{0} = 1$, and $\binom{m}{r} = 0$ when $m < r$ or $m < 0$.

Solution Denote the sum on the right-hand side by s_n. We can write

$$s_n = \sum_{j\geq 1}\binom{n-j}{j-1} = 1 + \sum_{i\geq 1}\binom{n-1-i}{i}.$$

Hence,

$$s_{n+1} - s_n = \sum_{i\geq 1}\left\{\binom{n-i}{i} - \binom{n-1-i}{i}\right\}.$$

Using the formula $\binom{m+1}{r} - \binom{m}{r} = \binom{m}{r-1}$, we get

$$s_{n+1} - s_n = \sum_{i\geq 1}\binom{n-1-i}{i-1} = s_{n-1}.$$

Thus, $s_{n+1} = s_n + s_{n-1}$. Now, $s_2 = 1$, $s_3 = 1 + 1$ and $s_4 = 1 + 2$. Thus, $s_n = u_n$ for $n = 2, 3$ and 4. If $s_n = u_n$ and $s_{n-1} = u_{n-1}$, then

$$s_{n+1} = s_n + s_{n-1} = u_n + u_{n-1} = u_{n+1}.$$

This proves the result by induction.

Set IX

1. Let $M_n = 2^n - 1$ be the nth Mersenne number. Suppose that p and $q = 2p + 1$ are primes. Prove that (i) q divides M_p if $q \equiv 1$ or $7 \pmod 8$, and (ii) q divides $M_p + 2$ if $q \equiv 3$ or $5 \pmod 8$. Deduce that M_{11} and M_{23} are not primes.

   Solution Suppose that $q \equiv 1$ or $7 \pmod 8$. Then, 2 is a quadratic residue of q, and hence $2^p = 2^{(q-1)/2} \equiv 1 \pmod q$; that is, $M_p = 2^p - 1$ is divisible by q. Similarly, if $q \equiv 3$ or $5 \pmod 8$, then 2 is a quadratic non-residue of q, and hence $2^p = 2^{(q-1)/2} \equiv -1 \pmod q$; that is, $M_p + 2 = 2^p + 1$ is divisible by q. This proves the first part.

   If $p = 11$, then $q = 23$ and if $p = 23$, then $q = 47$. In both cases, p and q are primes. Since $q \equiv 7 \pmod 8$, in either case, we see that q divides M_p. Thus, $23 | M_{11}$ and $47 | M_{23}$.

2. If p is an odd prime, prove that any prime divisor of M_p is of the form $2kp + 1$. Deduce that M_{13} is prime.

Solution Suppose that q is a prime divisor of M_p. Then, $2^p - 1 \equiv 0 \pmod{q}$. Let h be the smallest positive integer such that $2^h \equiv 1 \pmod{q}$. Then, p is divisible by h. Since p is prime, we must have $h = 1$ or p. If $h = 1$, then $2^h - 1 = 1$ which is impossible since $q|(2^h - 1)$. Hence, $h = p$. Further, $2^{q-1} \equiv 1 \pmod{q}$ by Fermat's theorem. Hence, $q - 1$ is also divisible by $h = p$. Thus, $q - 1$ is even and a multiple of p, that is, $q - 1 = 2kp$ for integer k. This gives the first statement.

Let q be the smallest prime divisor of M_{13}. Then, $q \le \sqrt{M_{13}} < 2^7$, and $q = 2kp + 1 = 26k + 1$. Now, $26k + 1 = 27, 53, 79, 105$ for $k = 1, 2,$ 3, 4, respectively. Of these, only 53 and 79 are primes. If $k \ge 5$, then $26k + 1 > 2^7$. Hence, the only possible choices for q are 53 and 79. Now, $2^{13} = 2(2^6)^2 \equiv 2(11)^2 \pmod{53}$. Thus,

$$M_{13} = 2^{13} - 1 \equiv 242 - 1 \equiv 29 \pmod{53}.$$

Similarly, $2^{13} \equiv 2(-15)^2 \pmod{79}$, and hence, $M_{13} \equiv 450 - 1 \equiv 54 \pmod{79}$. This shows that neither 53 nor 79 divides M_{13}; so M_{13} is prime.

3. If p is an odd prime, prove that any prime divisor q of M_p is of the form $q = 8m \pm 1$.

Solution Let q be a prime divisor of M_p. Then, $q = 2kp + 1$ for some integer k. Now,

$$q|M_p \implies 2^p \equiv 1 \pmod{q} \implies 2^{kp} \equiv 1 \pmod{q}.$$

This shows that $2^{(q-1)/2} = 2^{kp} \equiv 1 \pmod{q}$, that is, 2 is a quadratic residue of q. This is possible only if $q \equiv \pm 1 \pmod{8}$.

4. Prove that the Mersenne number M_{29} is composite by finding its smallest prime divisor.

Solution Let q be any prime divisor of M_{29}. Then, $q = 2kp + 1 = 58k + 1 = 59, 117, 175, 233$, and so on. By the previous problem, $q \equiv \pm 1 \pmod{8}$. Since $59 \equiv 3 \pmod{8}$, and 117 and 175 are not primes, the smallest possible choice of q is 233. Now, $2^8 = 256 \equiv 23 \pmod{233}$, and so

$$2^{14} \equiv 23 \times 64 \equiv 74 \pmod{233}, \quad 2^{28} \equiv 74^2 \equiv 117 \pmod{233}.$$

Thus, $2^{29} \equiv 234 \equiv 1 \pmod{233}$. This proves that $M_{29} = 2^{29} - 1$ is composite, the smallest prime dividing it being 233.

5. (a) Prove that any two Fermat numbers are relatively prime, that is, $\gcd(F_m, F_n) = 1$ for $m > n \geq 0$. Deduce that the number of primes is infinite.

   (b) Use the sequence $\{a_n\}$, defined by

$$a_1 = 2, \quad a_{n+1} = a_n^2 - a_n + 1, \quad \text{for } n \geq 1,$$

   to show that the number of primes is infinite.

Solution (a) If $m > n \geq 0$, let $m = n + k$. Then $2^m = 2^n \cdot 2^k$, and so

$$2^{2^m} = \left(2^{2^n}\right)^{2^k} = x^{2^k}, \quad \text{where } x = 2^{2^n}.$$

Thus, $F_m = x^{2^k} + 1$ and $F_n = x + 1$. But $x^{2^k} - 1$ is divisible by $x^2 - 1$, and hence by $x + 1$. It follows that $F_m - 2$ is divisible by F_n. Hence, if $d|F_m$ and $d|F_n$, then $d|2$. Note that d is odd since F_m and F_n are odd. This implies that if d is a common divisor of F_m and F_n, then $d = 1$, that is, $\gcd(F_m, F_n) = 1$.

Thus, no two Fermat numbers can have a common prime divisor. Since there are an infinite number of Fermat numbers, there must also exist an infinite number of primes.

(b) By definition of the sequence, we have $a_{n+1} - 1 = a_n(a_n - 1)$. Repeated application of this shows that

$$a_{n+1} - 1 = a_n a_{n-1} \cdots a_1(a_1 - 1) = a_n a_{n-1} \cdots a_1.$$

This shows that if p is a prime dividing a_{n+1}, then p cannot divide any a_i with $i \leq n$. Hence, as before, there must be an infinite number of primes.

6. Prove that if p is prime and $p > 3$, then,

$$3^{(p-1)/2} \equiv (-1)^{[(p+1)/6]} \pmod{p}.$$

Solution Let $m = [(p + 1)/6]$. Then,

$$m \leq \frac{p+1}{6} < m + 1, \quad 6m \leq p + 1 < 6m + 6.$$

Since p is prime and $p > 3$, we must have $p = 6m - 1$ or $6m + 1$.

Use Gauss' lemma. We have to consider the residues of the multiples of 3 having the least absolute value. Let $r_1, r_2, \ldots, r_{(p-1)/2}$ be given by

$$3i \equiv r_i \pmod{p}, \quad |r_i| < \frac{p}{2}, \quad 1 \le i \le \frac{p-1}{2}.$$

Clearly, $0 < 3i < 3p/2$ for $1 \le i \le (p-1)/2$, and

$$r_i = \begin{cases} 3i & \text{for } 0 < 3i < p/2 \\ 3i - p & \text{for } p/2 < 3i < 3p/2. \end{cases}$$

So $r_i < 0$ if and only if $p/2 < 3i < p$. If $p = 6m - 1$, then this holds if and only if $m \le i \le 2m - 1$. If $p = 6m + 1$, then this holds if and only if $m + 1 \le i \le 2m$. In either case, there are exactly m values of i for which $r_i < 0$. Hence,

$$3^{(p-1)/2} \equiv (-1)^m \pmod{p}$$

by Gauss' lemma.

7. Prove that, for $n \ge 1$, the Fermat number F_n is prime if and only if

$$3^{(F_n-1)/2} \equiv -1 \pmod{F_n}.$$

Solution First suppose that $p = F_n$ is prime. Then,

$$3^{(F_n-1)/2} = 3^{(p-1)/2} \equiv (-1)^{[(p+1)/6]} \pmod{p},$$

by Problem ??. Now, $p \equiv 1 \pmod{4}$, and

$$p \equiv (-1)^{2^n} + 1 \equiv 2 \pmod{3}.$$

Hence, $p = 12k + 5$ for some integer k, and $[(p+1)/6] = 2k + 1$. This shows that

$$3^{(F_n-1)/2} \equiv -1 \pmod{F_n}.$$

Conversely, suppose that $3^{(F_n-1)/2} \equiv -1 \pmod{F_n}$. Let p be any prime divisor of F_n. Since $3^{(F_n-1)/2} + 1$ is divisible by F_n and hence by p, we get

$$3^{(F_n-1)/2} \equiv -1 \pmod{p}, \quad 3^{F_n-1} \equiv 1 \pmod{p}.$$

Let h be the least positive integer such that $3^h \equiv 1 \pmod{p}$. Then, $h|(F_n - 1)$. Now, $F_n - 1 = 2^N$, where $N = 2^n$. Hence, $h = 2^r$ for some $r \leq N$. If $r \leq N - 1$, then $2^{N-1} = 2^r m = mh$ for some integer m. In that case, $\frac{1}{2}(F_n - 1) = 2^{N-1} = mh$, and

$$3^{(F_n-1)/2} = 3^{mh} \equiv 1 \pmod{p}.$$

If $3^{(F_n-1)/2} \equiv -1 \pmod{p}$, this gives a contradiction since p must be odd. It shows that $r = N$, $h = 2^N = F_n - 1$.

If $n \geq 1$, then $N = 2^n$ is even, and so $F_n - 1 = 2^N \equiv 1 \pmod{3}$. It follows that $p \neq 3$, and hence $3^{p-1} \equiv 1 \pmod{3}$ by Fermat's theorem. This implies that $h|(p-1)$, $F_n - 1 = h \leq p - 1$, and so $F_n \leq p$. But $p \leq F_n$, since $p|F_n$. Thus, $p = F_n$, proving that F_n is prime.

8. Any prime divisor p of the Fermat number $F_n = 2^{2^n} + 1$, where $n \geq 2$, is of the form $p = k \cdot 2^{n+2} + 1$. Deduce that $F_4 = 65537$ is prime.

Solution Let p be any prime divisor of $F_n = 2^{2^n} + 1$. Then, $2^{2^n} \equiv -1 \pmod{p}$. On squaring, we get $2^{2^{n+1}} \equiv 1 \pmod{p}$. Let h be the least positive integer such that $2^h \equiv 1 \pmod{p}$. Then $h|2^{n+1}$, and so $h = 2^r$ for some $r \leq n + 1$. If $r \leq n$, then $2^n = 2^r m = mh$ for some integer m. In that case, $2^{2^n} = 2^{mh} \equiv 1 \pmod{p}$, that is, $F_n \equiv 2 \pmod{p}$. This is not possible since p must be odd. This shows that $h = 2^{n+1}$.

Since $2^{p-1} \equiv 1 \pmod{p}$, we see that $p-1$ is divisible by $h = 2^{n+1}$ and so by 8 if $n > 1$. Thus, $p \equiv 1 \pmod{8}$ which implies that $2^{(p-1)/2} \equiv 1 \pmod{p}$. It follows that $h|(p-1)/2$. Thus, $p - 1$ is a multiple of $2h = 2^{n+2}$, proving the first statement.

Let p be the least prime divisor of F_4. By the first part, $p = 2^6 k + 1$, for some $k \geq 1$, that is, $p = 65, 129, 193, 257$, and so on. Note that if F_4 is not prime, then $p \leq \sqrt{F_4} = \sqrt{2^{16} + 1} < 2^8 + 1$. Hence, we need consider only $p = 193$. Now,

$$2^8 \equiv 63 \pmod{193}, \quad 2^{16} \equiv 3969 \equiv 109 \pmod{193}.$$

Hence, $F_4 = 2^{16} + 1$ is not divisible by 193. So F_4 is prime.

9. For $n \geq 1$, prove the following:

   (a) There are infinitely many composite numbers of the form $2^{2^n} + 3$.

   (b) Each of numbers $2^{2^n} + 5$ is composite.

Solution (a) We have $2^{2k+1} = 4^k \cdot 2 \equiv 2 \pmod 3$. Thus, $2^{2k+1} = 3m + 2$ for some integer m. Hence,

$$2^{2^{2k+1}} + 3 = 2^{3m+2} + 3 = 8^m \cdot 4 + 3 \equiv 0 \pmod 7.$$

This shows that $2^{2^n} + 3$ is divisible by 7 when n is odd.

(b) Since $2^{2k} + 5 \equiv 1 + 5 \equiv 0 \pmod 3$, we see that $2^{2^n} + 5$ is divisible by 3 for all n.

10. If $2^n \equiv 2 \pmod n$ and n is composite, then n is said to be a pseudoprime. Show that every Fermat number is either a prime or a pseudoprime.

Solution For $n \geq 0$, we have $2^n = n + k$ with $k > 0$. So $2^{2^n} = 2^n \cdot 2^k$; that is, $F_n - 1 = N2^n$, where $N = 2^k$. Thus,

$$2^{F_n - 1} = \left(2^{2^n}\right)^N = (F_n - 1)^N \equiv (-1)^N \pmod{F_n}.$$

Since N is even, we get $2^{F_n - 1} \equiv 1 \pmod{F_n}$, $2^{F_n} \equiv 2 \pmod{F_n}$. This proves the result.

11. If p and q are both primes and $q|(2^p + 1)$, prove that either $q = 3$ or $q = 2kp + 1$ for some integer k. Use this to determine the smallest prime divisor $q > 3$ of $2^{29} + 1$.

Solution If $q|(2^p + 1)$, then $2^p \equiv -1 \pmod q$, and $2^{2p} \equiv 1 \pmod q$. Let h be the smallest positive integer such that $2^h \equiv 1 \pmod q$. Then, $h|(2p)$, and so $h = 1, 2, p$ or $2p$. Now, $h \neq 1$ as $q \neq 1$. If $h = 2$, then $q = 3$; and if $h = p$, then $2^p = 2^h \equiv 1 \pmod q$. This is not possible since $2^p \equiv -1 \pmod q$, and $q \neq 2$. It follows that if $q \neq 3$, then $h = 2p$. Since $2^{q-1} \equiv 1 \pmod q$ by Fermat's theorem, we see that $q - 1$ is divisible by $h = 2p$, that is, $q - 1 = 2kp$.

Let $p = 29$ and q be a prime dividing $2^p + 1$. If $q > 3$, then $q = 58k + 1$ for some k. The least such q is 59. Modulo 59 we have

$$2^7 \equiv 10, \quad 2^{14} \equiv -18, \quad 2^{28} \equiv -30, \quad 2^{29} \equiv -1.$$

Thus, the smallest prime divisor $q > 3$ of $2^{29} + 1$ is 59.

12. If p is a prime, prove that the order of 2 modulo p is a power of 2 if and only if p divides a Fermat number, F_k. Also show that there is no Fermat number that is divisible by 7, 13 or 97.

Solution Let h be of the order of 2 modulo p. Then, h is the least positive integer such that $2^h \equiv 1 \pmod{p}$. Using the identity

$$2^{2^n} - 1 = \left(2^{2^{n-1}} - 1\right)\left(2^{2^{n-1}} + 1\right),$$

we see that $F_n - 2 = (F_{n-1} - 2)F_{n-1}$. It follows that

$$F_n - 2 = F_0 F_1 F_2 \cdots F_{n-1}.$$

First suppose that p divides F_k for some $k \le n-1$. Then p divides $F_n - 2$, and hence, $2^{2^n} - 1 = F_n - 2 \equiv 0 \pmod{p}$. This implies that $h|2^n$ so that $h = 2^r$ for some r with $0 \le r \le n$.

Conversely, suppose that $h = 2^r$ for some r. Since p divides $2^h - 1$ and

$$2^h - 1 = 2^{2^r} - 1 = F_r - 2 = F_0 F_1 F_2 \cdots F_{r-1},$$

we see that $p|F_k$ for some $k \le r - 1 \le n - 1$.

Since $2^{p-1} \equiv 1 \pmod{p}$, we see that if h is the order of 2 modulo p, then $h|p-1$. If $p = 7$, then $h = 3$. If $p = 13$, then $h|12$. Clearly, $h \ne 2$ and $h \ne 4$. Let $p = 97$. Then, $h|96$. It is easily seen that h cannot be equal to 2, 4, 8, 16 or 32. Thus, h is not a power of 2 in any of these cases.

13. For any prime $p > 3$, prove that $(2^p + 1)/3$ has a prime divisor greater than p.

Solution Let $N = (2^p + 1)/3$. Then,

$$N = \frac{2^p + 1}{2 + 1} = 2^{p-1} - 2^{p-2} + \cdots - 2 + 1 = \sum_{r=1}^{p} (-2)^{p-r}.$$

Since $(-2)^{p-r} \equiv 1^{p-r} \pmod{3}$, each term in the sum for N is congruent to 1 modulo 3. Thus, $N \equiv p \pmod{3}$; so N is not divisible by 3 since $p > 3$. Let q be any prime divisor of N. Then, q divides $3N = 2^p + 1$. By Problem ??, we see that $q = 2kp + 1$ for some integer k. Hence, $q > p$.

14. Let p be an odd prime. Prove that

 (a) p can be expressed as the difference of two squares,

 (b) p can be expressed as the difference of two cubes if and only if it is of the form $6t_n + 1$, where t_n is a triangular number,

(c) any Fermat prime can be expressed as the difference of two squares, but not of two cubes.

Solution (a) If $x = (p+1)/2$ and $y = (p-1)/2$, then $x^2 - y^2 = p$.

(b) Suppose that $p = x^3 - y^3$, where x and y are positive integers. Then, $p = (x-y)(x^2+xy+y^2)$. Since, p is prime and $x^2+xy+y^2 > 1$, we must have $x - y = 1$. Hence, if $y = n$, then $x = n+1$ and

$$p = (n+1)^3 - n^3 = 3n^2 + 3n + 1 = 6t_n + 1.$$

On the other hand, $6t_n + 1 = 3n^2 + 3n + 1 = (n+1)^3 - n^3$. This proves (b).

(c) If N is odd, then $N = \left(\frac{1}{2}(N+1)\right)^2 - \left(\frac{1}{2}(N-1)\right)^2$. Consequently, every Fermat number is the difference of two squares.

If a Fermat prime F_n can be expressed as the difference of two cubes, then $F_n - 1$ must be divisible by 6, using (b). This is not true since $F_n - 1 = 2^{2^n}$.

15. (a) If u and v are real numbers, and m is a positive integer, prove that

$$u^{2m} + u^{2m-1}v + u^{2m-2}v^2 + \cdots + v^{2m} \geq \frac{1}{2}u^{2m}.$$

(b) If p and q are odd primes and if $q = x^p - y^p$, where x, y are integers, prove that $q \equiv 1 \pmod{p}$.

Solution (a) The inequality is trivial if u or v is zero. So let u, v be non-zero, and $t = v/u$. The inequality reduces to

$$1 + t + t^2 + \cdots + t^{2m} \geq \frac{1}{2}, \quad 1 + 2t + 2t^2 + \cdots + 2t^{2m} \geq 0,$$

that is,

$$(1+2t+t^2) + t^2(1+2t+t^2) + \cdots + t^{2m-2}(1+2t+t^2) + t^{2m} \geq 0.$$

This is obviously true.

(b) If $q = x^p - y^p$, then $q = (x-y)z$, where

$$z = x^{p-1} + x^{p-2}y + \cdots + y^{p-1}.$$

From (a), we see that $z \geq \frac{1}{2}x^{p-1}$ and $z \geq \frac{1}{2}y^{p-1}$. Thus, z is a positive divisor of q; hence, $z = 1$ or $z = q$ since q is prime. If $z = 1$, then

$|x| \le 1$ and $|y| \le 1$. Since $q = x^p - y^p$ is prime, we cannot have $x = 0$ or $y = 0$. Further, $|x| = 1 = |y|$ only when $q = 2$. So we cannot have $z = 1$ if q is an odd prime. This shows that $z = q$ and $x - y = 1$. Since $x^p - y^p \equiv x - y \pmod{p}$, we get the desired result.

16. If a Fermat prime F_n can be expressed as $F_n = x^k - y^k$, where x, y are integers, prove that $k = 2$.

Solution Suppose that k has an odd prime divisor p. Let $k = pm$, $u = x^m$ and $v = y^m$. Then $F_n = u^p - v^p$. If F_n is prime, this implies that $F_n - 1$ is divisible by p by Problem **??**(b). Since this is not possible, we see that k has no odd prime divisor. Consequently, either $k = 2$ or k is divisible by 4. If k is divisible by 4, let $a = x^{k/4}$ and $b = y^{k/4}$. Then, $F_n = a^4 - b^4 = (a^2 - b^2)(a^2 + b^2)$. Since F_n is prime, this implies that $a^2 - b^2 = 1$ or $a^2 + b^2 = 1$. In the former case, $a + b = a - b = \pm 1$. So, in any case, $a = 1$, $b = 0$, giving a contradiction. Hence, we must have $k = 2$.

17. Prove that a Carmichael number is necessarily odd.

Solution If n is a Carmichael number, then $(n-1)^n \equiv n - 1 \equiv -1 \pmod{n}$. Since $(n-1)^n \equiv (-1)^n \pmod{n}$, we get $(-1)^n \equiv -1 \pmod{n}$. This is possible only if n is odd or $n = 2$. Since a Carmichael number is composite, it must be odd.

18. (a) Let $n = 2p$, where p is an odd prime. Prove that $a^{n-1} \equiv a \pmod{n}$ for any a.

(b) Does there exist an odd number n with the property that $a^{n-1} \equiv a \pmod{n}$ for every a?

(c) If $n = 195$, prove that $a^{n-2} \equiv a \pmod{n}$ for any a.

Solution (a) If $(a, p) = 1$, then $a^{n-2} = a^{2(p-1)} \equiv 1 \pmod{p}$. Hence, $a^{n-1} - a = a(a^{n-2} - 1)$ is divisible by p for all a. If a is odd, then $a^{n-2} - 1$ is even. Hence, $a^{n-1} - a = a(a^{n-2} - 1)$ is divisible by 2 for all a. Thus, $a^{n-1} - a$ is divisible by $2p = n$, for any a.

(b) Suppose that $a^{n-1} \equiv a \pmod{n}$ for every a. Then, $(n-1)^{n-1} \equiv n - 1 \equiv -1 \pmod{n}$. Since $(n-1)^{n-1} \equiv (-1)^{n-1} \pmod{n}$, we get $(-1)^{n-1} \equiv -1 \pmod{n}$. This is possible only if n is even. So the answer to the question is 'No'.

(c) Note that $n = 195 = 3 \cdot 5 \cdot 13$. Let $(a, n) = d$, $a = bd$ and $n = md$. Then, $d = 1, p, pq$ or $pqr = n$, where p, q, r denote the primes 3, 5, 13 in some order. If $d = p$, them $m = qr$, and if $d = pq$, then $m = r$.

Thus, we can see that $\gcd(a, m) = 1$ in each of these cases. Hence, $a^{\phi(m)} \equiv 1 \pmod{m}$. Moreover, $\phi(n) = (p-1)(q-1)(r-1)$, which is divisible by $\phi(qr)$ and $\phi(r)$. Thus, $\phi(m)|\phi(n)$. Since $\phi(n) = 96$, we see that $192 = n - 3$ is divisible by $\phi(m)$. This shows that $a^{n-3} \equiv 1 \pmod{m}$. Hence,

$$a^{n-2} - a = a(a^{n-3} - 1) = bd(a^{n-3} - 1).$$

This is divisible by $md = n$, and hence $a^{n-2} \equiv a \pmod{n}$.

19. Show that 341 and 645 are pseudo-primes, but not Carmichael numbers.

Solution First observe that $341 = 11 \cdot 31$ and $645 = 3 \cdot 5 \cdot 43$ are composite. Since $2^{10} \equiv 1 \pmod{11}$, we get $2^{340} \equiv 1 \pmod{11}$. As $2^5 \equiv 1 \pmod{31}$, we get $2^{340} \equiv 1 \pmod{31}$. This shows that $2^{340} - 1$ is divisible by 11 and 31, and hence by 341. Thus, $2^{341} \equiv 2 \pmod{341}$, and hence 341 is a pseudo-prime.

Similarly, since $2^2 \equiv 1 \pmod{3}$ and $2^4 \equiv 1 \pmod{5}$, we get $2^{644} \equiv 1 \pmod{3}$ and $2^{644} \equiv 1 \pmod{5}$. Further, $2^7 \equiv -1 \pmod{43}$, $2^{14} \equiv 1 \pmod{43}$, and so $2^{644} \equiv 1 \pmod{43}$. It follows that $2^{644} - 1$ is divisible by 3, 5 and 43, and hence by 645. Thus, $2^{645} \equiv 2 \pmod{645}$, and hence 645 is a pseudo-prime.

Since $3^{30} \equiv 1 \pmod{31}$, we get $3^{330} \equiv 1 \pmod{31}$. Further, $3^5 \equiv -5 \pmod{31}$, $3^{10} \equiv 25 \pmod{31}$, and so $3^{340} \equiv 25 \pmod{31}$. Thus, $3^{340} - 1$ is not divisible by 31, and hence by 341. This shows that 341 is not a Carmichael number.

Similarly, $7^3 = 343 \equiv -1 \pmod{43}$, and hence, $7^{645} \equiv (-1)^{215} \equiv -1 \pmod{43}$. This shows that 645 is not a Carmichael number.

20. Show that $2 \cdot 73 \cdot 1103$ is an even pseudo-prime.

Solution Let $n = 2 \cdot 73 \cdot 1103$. We have $2^{72} \equiv 1 \pmod{73}$ and

$$n = 73 \cdot 2206 \equiv 2206 \equiv 46 \pmod{72}.$$

If $n = 72m + 46$, then

$$2^{n-1} = 2^{72m+45} \equiv 2^{45} \pmod{73}.$$

Since $2^9 = 512 \equiv 1 \pmod{73}$, we get $2^{n-1} \equiv 1 \pmod{73}$.

Similarly, $2^{1102} \equiv 1 \pmod{1103}$ and

$$n = 1103 \cdot 146 \equiv 146 \pmod{1102}.$$

If $n = 1102m_1 + 146$, then

$$2^{n-1} = 2^{1102m_1+145} \equiv 2^{145} \pmod{1103}.$$

Modulo 1103 we have,

$$2^{10} \equiv -79, \quad 2^{14} \equiv -161, \quad 2^{28} \equiv 552, \quad 2^{29} \equiv 1.$$

Hence, $2^{145} \equiv 1 \pmod{1103}$. Thus, we see that $2^{n-1} - 1$ is divisible by 73 and 1103. Consequently, $2^n - 2 = 2(2^{n-1} - 1)$ is divisible by n. So n is a pseudo-prime.

21. Show that there exist an infinite number of pseudo-primes.

Solution Let $M_n = 2^n - 1$. If n is composite so is M_n. Suppose that n is a pseudo-prime. Then $M_n - 1 = 2^n - 2$ is divisible by n. Let $M_n - 1 = kn$. Then

$$2^{M_n - 1} = (2^n)^k = (M_n + 1)^k \equiv 1 \pmod{M_n}.$$

Hence, $2^{M_n} \equiv 2 \pmod{M_n}$. Thus, M_n is a pseudo-prime if n is a pseudo-prime. This implies that given any pseudo-prime n we can find a larger pseudo-prime M_n. This proves the result.

22. (a) Let p be a prime. Prove that p^2 is a pseudo-prime if and only if $2^{p-1} \equiv 1 \pmod{p^2}$.

(b) Show that there are positive integers n such that n^2 is a pseudo-prime.

Solution (a) If $p = 2$, then $p^2 = 4$ is not a pseudo-prime, and $2^{p-1} \not\equiv 1 \pmod{p^2}$. So let p be an odd prime.

Since $2^{p-1} \equiv 1 \pmod{p}$ and $2^{\phi(p^2)} \equiv 1 \pmod{p^2}$, we have

$$2^{p-1} = 1 + ap, \quad 2^{p^2-p} = 2^{\phi(p^2)} = 1 + bp^2,$$

for some integers a, b. Hence,

$$2^{p^2-1} = 2^{(p^2-p)+(p-1)} = (1 + ap)(1 + bp^2) \equiv 1 + ap \pmod{p^2}.$$

Thus, $2^{p^2-1} \equiv 1 \pmod{p^2}$ if and only if a is divisible by p. This proves (a).

(b) Let n be a prime such that $2^{n-1} \equiv 1 \pmod{n^2}$. Then, n^2 is a pseudo-prime by (a). There are at least two primes n, namely, 1093 and 3511, such that $2^{n-1} \equiv 1 \pmod{n^2}$. This gives (b).

23. (a) Let $a > 1$ and m be a composite number such that $a^m \equiv a \pmod{m}$. If m is relatively prime to $a-1$, show that there exist an infinite number of composite numbers n such that $a^n \equiv a \pmod{n}$.

(b) Show that there exist an infinite number of composite numbers n such that $3^n \equiv 3 \pmod{n}$.

Solution (a) Note that $a^m - a = a(a^{m-1} - 1)$ is divisible by $a-1$. So if it is also divisible by m and $\gcd(m, a-1) = 1$, then it is divisible by $m(a-1)$. Let $a^m - a = km(a-1)$ and

$$n = \frac{a^m - 1}{a - 1} = 1 + a + a^2 + \cdots + a^{m-1}.$$

Then $a^m = (a-1)n + 1 \equiv 1 \pmod{n}$, and

$$(km+1)(a-1) = a^m - a + a - 1 = n(a-1), \quad km + 1 = n.$$

Hence,

$$a^{n-1} = (a^m)^k \equiv 1 \pmod{n}, \quad a^n \equiv a \pmod{n}.$$

Further,

$$n - m = (a-1) + (a^2 - 1) + \cdots + (a^{m-1} - 1) \equiv 0 \pmod{a-1}.$$

Hence, $\gcd(n, a-1) = 1$ if $\gcd(m, a-1) = 1$. This shows that n has the same properties as m. Since $n > m$ and the argument can be repeated any number of times with m replaced by n, we see that there exist an infinite number of integers n such that $a^n \equiv a \pmod{n}$.

(b) Let m be a Carmichael number, and $a = 3$. Then, $a^m \equiv a \pmod{m}$. Since m is necessarily odd and $a - 1 = 2$, we see that $\gcd(m, a-1) = 1$. So the result follows from (a).

24. Let $n = p_1 p_2 \cdots p_k$, where the p_i are distinct primes and $k > 1$. If $p_i - 1$ divides $n - 1$ for each i, prove that $a^n \equiv a \pmod{n}$ for all positive integers a.

Solution First note that n is composite since $k > 1$. Let a be a positive integer, and

$$\gcd(a, n) = d, \quad a = bd, \quad n = md.$$

Then $d = 1, p_i$ or a product of some of the primes p_i. Since n is a product of distinct primes we see that if $m > 1$, then $m = n/d$ is the product of those primes which do not divide d. It follows that if $p_i | m$, then p_i does not divide d and so p_i cannot divide a. Hence, $a^{p_i - 1} \equiv 1 \pmod{p_i}$ by Fermat's theorem. It is given that $n - 1$ is divisible by $p_i - 1$. This implies that $a^{n-1} \equiv 1 \pmod{p_i}$. Hence, $a^{n-1} - 1$ is divisible by each p_i that divides m. Consequently, $a^{n-1} - 1$ is divisible by m. This is trivial when $m = 1$. Thus,

$$a^n - a = a(a^{n-1} - 1) = bd(a^{n-1} - 1);$$

and this is divisible by $md = n$. Hence, $a^n \equiv a \pmod{n}$.

25. Suppose that n is a composite number such that $a^{n-1} \equiv 1 \pmod{n}$ whenever a is relatively prime to n. Prove that n is product of distinct primes.

Solution Let $p_1, \ldots, p_k$ be the distinct prime divisors of n, and $P = p_1 \cdots p_k$ be their product. Suppose that n is divisible by p_1^2.

Now, $1 + P$ is relatively prime to n, and so $(1 + P)^{n-1} - 1$ is divisible by n. But

$$\begin{aligned}(1 + P)^{n-1} - 1 &= \binom{n-1}{1} P + \binom{n-1}{2} P^2 + \cdots \\ &\equiv (n-1)P \pmod{p_1^2}.\end{aligned}$$

Since $(1 + P)^{n-1} - 1$ is divisible by n, and n is divisible by p_1^2, we see that

$$(n-1)p_1 \cdots p_k \equiv 0 \pmod{p_1^2}.$$

If $k = 1$, this implies that $(n - 1) \equiv 0 \pmod{p_1}$, and if $k > 1$, it implies that $(n-1)p_2 \cdots p_k \equiv 0 \pmod{p_1}$. Since $n-1$ is not divisible by p_1 and the p_i are distinct, this gives a contradiction. It follows that n cannot be divisible by p_1^2. Similarly, n cannot be divisible by the square of any p_i. This proves the result.

26. Let $n = p_1 p_2 \cdots p_k$, where the p_i are distinct primes. Suppose also that $a^{n-1} \equiv 1 \pmod{n}$ whenever a is relatively prime to n. Prove that $p_i - 1$ divides $n - 1$ for each i.

Solution Consider a fixed factor p_i, and let $n = mp_i$. Let g be a primitive root of p_i. Then any number, congruent to g modulo p_i, is also a primitive root of p_i. Since m is relatively prime to p_i, the congruence

$$p_i x \equiv 1 - g \pmod{m}$$

has a solution, that is, there exist integers x and y such that $p_i x = 1 - g + my$. Let $a = g + p_i x$. Then, $a \equiv g \pmod{p_i}$ and $a = 1 + my$. Thus, a is relatively prime to p_i and also to m, and hence, to $mp_i = n$. So $a^{n-1} \equiv 1 \pmod{n}$.

Let $n - 1 = q(p_i - 1) + r$, where $0 \le r < p_i - 1$. Then,

$$a^{n-1} = a^{q(p_i-1)+r} = \left(a^{p_i-1}\right)^q a^r. \qquad (*)$$

Now, $a^{p_i-1} \equiv 1 \pmod{p_i}$; and $a^{n-1} - 1$ is divisible by n and hence by p_i. It follows from (*) that $a^r \equiv 1 \pmod{p_i}$. Since a is a primitive root of p_i and $r < p_i - 1$, this is possible only if $r = 0$, in which case, $n - 1$ is divisible by $p_i - 1$. This holds for each prime factor of n.

27. Prove that for a composite number n, the following two statements are equivalent:

 (i) $a^n \equiv a \pmod{n}$ for all positive integers a;

 (ii) $a^{n-1} \equiv 1 \pmod{n}$ for all positive integers a which are relatively prime to n.

 Solution Suppose that n is a composite number satisfying (i). To prove (ii), let $\gcd(a, n) = 1$. Then, by cancellation law, $a^{n-1} \equiv 1 \pmod{n}$. This gives (ii).

 Conversely suppose that n is a composite number satisfying (ii). Then, by Problem ??, n is a product of distinct primes $p_1, \ldots, p_k$. Moreover, $p_i - 1$ divides $n - 1$ for each i by Problem ??. So Problem ?? shows that (i) holds.

28. Prove that 561, 1105, 1729 and 2465 are Carmichael numbers.

 Solution Apply Problem ??. If $n = 561$, then $n = 3 \cdot 11 \cdot 17$, and $n - 1 = 560$ is divisible by 2, 10 and 16. So $a^{561} \equiv a$ for all a; that is, 561 is a Carmichael number.

 Similarly, if $n = 1105$, then $n = 5 \cdot 13 \cdot 17$, and $n - 1 = 1104$ is divisible by 4, 12 and 16. If $n = 1729$, then $n = 7 \cdot 13 \cdot 19$, and $n - 1 = 1728$ is divisible by 6, 12 and 18. If $n = 2465$, then $n = 5 \cdot 17 \cdot 29$, and $n - 1 = 2464$ is divisible by 4, 16 and 28. This completes the verification.

29. Prove that if $6k + 1$, $12k + 1$ and $18k + 1$ are primes, then

$$n = (6k + 1)(12k + 1)(18k + 1)$$

is a Carmichael number. Verify that 1729 is number of this form.

Solution Writing $x = 6k$ in the continued product for n, we get

$$n = 6x^3 + 11x^2 + 6x + 1 = 6(216k^3) + 11(36k^2) + 6(6k) + 1.$$

This shows that $n - 1$ is divisible by $6k$, $12k$ and $18k$. So an application of Problem **??** shows that n is a Carmichael number.

If $k = 1$, then $n = 7 \cdot 13 \cdot 19 = 1729$.

30. Suppose that n is a composite number such that $a^{n-1} \equiv 1 \pmod{n}$ whenever a is relatively prime to n. Prove that n is a product of at least three primes.

    Solution Problem **??** shows that n is a product of distinct primes. Suppose that it is the product of two primes p and q with $p > q$. Let g be a primitive root of p. If g is divisible by q, then $g + p$ cannot be divisible by q. Let $a = g$ or $g + p$ so that $(a, q) = 1$. Then, a is a primitive root of p. Further, a is relatively prime to p and also to q; hence, it is relatively prime to $n = pq$, which implies that $a^{n-1} \equiv 1 \pmod{n}$. Thus, $a^{n-1} - 1$ is divisible by $n = pq$ and hence by p. But

    $$a^{n-1} = a^{q(p-1)} \cdot a^{q-1} \equiv a^{q-1} \pmod{p}.$$

    It follows that $a^{q-1} \equiv 1 \pmod{p}$. Since $0 < q - 1 < p - 1$ and a is a primitive root of p, we get a contradiction. So n must be a product of at least three primes.

Set X

1. Recall that an integer $n > 1$ is highly composite if it has more divisors than all positive integers less than it. Any $n > 1$ can be expressed as

    $$n = 2^{n_1} 3^{n_2} 5^{n_3} \cdots p_r^{n_r},$$

    where p_i is the ith prime, $n_i \geq 0$ and $n_r > 0$. Prove that if n is highly composite, then $n_1 \geq n_2 \geq \cdots \geq n_r$. Using Bertrand's theorem, or otherwise, show that the last exponent n_r is either 1 or 2. Determine those highly composite numbers for which $n_r = 2$.

Solution Suppose that $n_i < n_{i+1}$. We can write $n = p_i^{n_i} p_{i+1}^{n_{i+1}} N$. Let $m = p_i^{n_{i+1}} p_{i+1}^{n_i} N$. Since $p_i < p_{i+1}$, we have

$$p_i^{n_{i+1}-n_i} < p_{i+1}^{n_{i+1}-n_i},$$

and hence,

$$p_i^{n_{i+1}} p_{i+1}^{n_i} < p_i^{n_i} p_{i+1}^{n_{i+1}}.$$

This implies that $m < n$. The number of divisors of n and m are

$$d(n) = (n_i + 1)(n_{i+1} + 1)d(N), \quad d(m) = (n_{i+1} + 1)(n_i + 1)d(N).$$

So $d(m) = d(n)$. This is not possible if n is highly composite. It follows that $n_i \geq n_{i+1}$ for all i. This proves the first statement.

To prove that $n_r = 1$ or 2: We shall first prove that if $r > 1$ and $n_r > 1$, then $n_1 = n_r = 2$. We can write $n = 2^{n_1} p_r^{n_r} N$. Let $m = 2^{n_1 - 1} p_r^{n_r - 1} p_{r+1} N$. Since $p_{r+1} < 2p_r$, by Bertrand's theorem, we see that $m < n$. So $d(m) < d(n)$ as n is highly composite. Since $d(n) = (n_1 + 1)(n_r + 1)d(N)$ and $d(m) = n_1 n_r 2d(N)$, we get

$$2n_1 n_r < (n_1 + 1)(n_r + 1) = n_1 n_r + n_1 + n_r + 1,$$

$$(n_1 - 1)(n_r - 1) < 2.$$

So we must have $n_1 - 1 = n_r - 1 = 1$, that is, $n_1 = n_r = 2$.

On the other hand, if $r = 1$, then $n = 2^{n_1}$. If $n_1 > 2$, let $m = 2^{n_1 - 2}3$. Then, $d(n) = n_1 + 1$ and $d(m) = (n_1 - 1)2$. So $d(m) \geq d(n)$. This is not possible as $m < n$ and n is highly composite. Hence $n_r = n_1 = 1$ or 2.

It remains to find all those highly composite numbers for which $n_r = 2$. If $r = 1$, then $n = 2^2$, and if $r = 2$, then $n_1 = n_2 = 2$ and $n = 2^2 \cdot 3^2$. These two are highly composite.

We proved above that if $r > 1$ and $n_r > 1$, then $n_1 = n_r = 2$, and so $n_i = 2$ for all i. If $r = 3$, then $n = 2^2 \cdot 3^2 \cdot 5^2$. If $m = 2^3 \cdot 3 \cdot 5 \cdot 7$, then $d(n) = 3^3$ and $d(m) = 4 \cdot 2^3$. So n cannot be highly composite as $m < n$ and $d(m) > d(n)$. Finally, let $r \geq 4$,

$$n = 2^2 5^2 p_r^2 N, \quad m = 2^3 (5p_r) p_{r+1} N.$$

Then, $d(n) = 3^3 d(N)$ and $d(m) = 4 \cdot 2^3 d(N)$. Hence, $d(n) < d(m)$. Since $p_{r+1} < 2p_r$, we have $2p_{r+1} < 4p_r < 5p_r$. This implies that $m < n$. So n cannot be highly composite if $r \geq 4$. Thus, we get only two highly composite numbers with $n_r = 2$, namely, 2^2 and $2^2 \cdot 3^2$.

2. Let $\gcd(a, b) = 1$ and c be a positive integer. Prove that there is an integer x such that $\gcd(a + bx, c) = 1$.

Solution If $c = 1$, there is nothing to prove. Otherwise, let $p_1, \ldots, p_k$ be the distinct prime divisors of c. Some of them may divide a or b, but none divides both a and b since $\gcd(a, b) = 1$. Suppose that $p_1, \ldots, p_r$ divide a, and $p_{r+1}, \ldots, p_s$ divide b. Let $m = p_{s+1} \cdots p_k$, the product of those primes which divide neither a nor b. If $s = k$, we take $m = 1$. Consider any prime divisor p_i of c. If $i \leq r$, then p_i divides a while it does not divide b or m. If $i > r$, then p_i does not divide a, and it divides b or m. In either case, p_i cannot divide $a + bm$. Thus, we see that there is no prime dividing c and $a + bm$. It follows that $a + bm$ is relatively prime to c.

3. Suppose that $p = 12k + 1$ and $q = 24k + 1$ are primes for some natural number k. If $n = pq$, prove that $2^n \equiv 2 \pmod{n}$ and $3^n \equiv 3 \pmod{n}$.

Solution Let $a = 2$ or 3. In either case, a is relatively prime to both $p = 12k + 1$ and $q = 24k + 1$. So by Fermat's theorem, $a^{p-1} \equiv 1 \pmod{p}$ and $a^{q-1} \equiv 1 \pmod{q}$. Now,

$$n - 1 = pq - 1 = p(q - 1) + (p - 1) = (p - 1)q + (q - 1).$$

It follows that

$$a^{n-1} \equiv a^{p-1} \pmod{q}, \quad a^{n-1} \equiv a^{q-1} \pmod{p}. \tag{1}$$

First observe that $q - 1 = 24k = 2(p - 1)$. Thus, $a^{q-1} \equiv 1 \pmod{p}$, and hence from (1)

$$a^{n-1} \equiv 1 \pmod{p}. \tag{2}$$

It is known that, modulo q, we have

$$2^{(q-1)/2} \equiv (-1)^{[(q+1)/4]} \quad \text{and} \quad 3^{(q-1)/2} \equiv (-1)^{[(q+1)/6]}.$$

But $[(q+1)/4] = \left[6k + \frac{1}{2}\right] = 6k$ and $[(q+1)/6] = \left[4k + \frac{1}{3}\right] = 4k$. Hence,

$$(-1)^{[(q+1)/4]} = 1 = (-1)^{[(q+1)/6]}.$$

Since $(q - 1)/2 = p - 1$, we get

$$2^{p-1} \equiv 1 \pmod{q} \quad \text{and} \quad 3^{p-1} \equiv 1 \pmod{q}.$$

Thus, $a^{p-1} \equiv 1 \pmod{q}$ whether $a = 2$ or 3. It follows from (1) that

$$a^{n-1} \equiv a^{p-1} \equiv 1 \pmod{q}.$$

Combining this with (2), we see that $a^{n-1} - 1$ is divisible by p and q, and hence by $pq = n$. This implies that $a^n \equiv a \pmod{n}$ if $a = 2$ or 3.

4. (a) Let p be an odd prime and t be a positive integer not divisible by p. Prove that if p can be represented as $p = x^2 + ty^2$ with positive integers x and y, then $(-t)^{(p-1)/2} \equiv 1 \pmod{p}$, that is, $-t$ is a quadratic residue of p.

   (b) Prove that a prime p can be expressed as $p = a^2 + 2b^2$ for some positive integers a and b, if and only if p is of the form $8k + 1$ or $8k + 3$.

Solution (a) Let $p = x^2 + ty^2$. Then $x^2 \equiv -ty^2 \pmod{p}$, and hence,

$$x^{p-1} \equiv (-t)^{(p-1)/2} y^{p-1} \pmod{p}.$$

Since p does not divide t, we see that $p|x$ if and only if $p|y$. If both x and y are divisible by p, then $x^2 + ty^2$ is divisible by p^2, which is not possible. This shows that neither x nor y is divisible by p, and so,

$$x^{p-1} \equiv 1 \equiv y^{p-1} \pmod{p}.$$

This readily show that $(-t)^{(p-1)/2} \equiv 1 \pmod{p}$.

(b) If p can be expressed as $p = a^2 + 2b^2$, then by (a),

$$(-2)^{(p-1)/2} \equiv 1 \pmod{p}. \qquad (*)$$

Suppose, on the other hand, that (*) holds. Then -2 is a quadratic residue of p, that is, there is an integer t, relatively prime to p, such that $t^2 \equiv -2 \pmod{p}$. By an earlier problem [Set III, Problem ??] there exist integers x and y such that

$$tx \equiv y \pmod{p}, \quad 0 < |x| < \sqrt{p}, 0 < |y| < \sqrt{p}.$$

Hence,

$$y^2 \equiv t^2x^2 \equiv -2x^2 \pmod{p}, \quad 0 < x^2 < p, 0 < y^2 < p.$$

Thus, $2x^2 + y^2 \equiv 0 \pmod{p}$ and $0 < 2x^2 + y^2 < 3p$. Hence, $2x^2 + y^2 = p$ or $2p$. If $2x^2 + y^2 = 2p$, then y^2 is even. So $y = 2z$, and

$x^2 + 2z^2 = p$ in this case. Thus, we have either $2x^2 + y^2 = p$, or else, $x^2 + 2z^2 = p$; in either case, p is of the form $a^2 + 2b^2$.

We have proved that p is expressible as $p = a^2 + 2b^2$ if and only if (*) holds. Since

$$2^{(p-1)/2} \equiv (-1)^{[(p+1)/4]} \pmod{p},$$

we see that (*) holds if and only if $(p-1)/2 + [(p+1)/4]$ is even. If p is any odd prime, then it is of the form $8k+1$, $8k+3$, $8k+5$, or $8k+7$. In these cases we have respectively,

$$(p-1)/2 + [(p+1)/4] = 4k + 2k, (4k+1) + (2k+1),$$
$$(4k+2) + (2k+1), (4k+3) + (2k+2).$$

Thus, $(p-1)/2 + [(p+1)/4]$ is even when $p = 8k+1$ or $8k+3$, and is odd in the other cases. This completes the proof of (b).

5. (a) Prove that a prime p can be represented as $p = a^2 + 3b^2$ with positive integers a and b, if and only if it is of the form $6k+1$.
   (b) Prove that a prime p can be simultaneously expressed in the forms $x^2 + y^2$, $u^2 + 2v^2$, and $r^2 + 3s^2$, if and only if $p = 24k + 1$ for some integer k.

Solution (a) Proceed as in the previous problem. Let $p = a^2 + 3b^2$ with positive integers a, b. Then $p > 3$. Hence,

$$(-3)^{(p-1)/2} \equiv 1 \pmod{p}, \qquad (*)$$

by (a) of the previous problem. Suppose, on the other hand, that (*) holds. Then -3 is a quadratic residue of p, that is, there is an integer t, relatively prime to p, such that $t^2 \equiv -3 \pmod{p}$. Hence, as before, there exist integers x and y such that

$$y^2 \equiv t^2x^2 \equiv -3x^2 \pmod{p}, \quad 0 < x^2 < p, 0 < y^2 < p.$$

Thus, $3x^2 + y^2 \equiv 0 \pmod{p}$ and $0 < 3x^2 + y^2 < 4p$. Hence, $3x^2 + y^2 = p$, $2p$ or $3p$. Note that

$$3x^2 \equiv 0 \text{ or } 3 \pmod{4}, \quad y^2 \equiv 0 \text{ or } 1 \pmod{4}.$$

Hence, $3x^2 + y^2 \equiv 0+0, 0+1, 3+0$ or $3+1 \pmod{4}$. But $2p \equiv 2 \pmod{4}$ for any odd prime p. So we rule out the possibility $3x^2 +$

$y^2 = 2p$. If $3x^2 + y^2 = 3p$, then y is divisible by 3. So $y = 3z$, and $x^2 + 3z^2 = p$ in this case. Thus we have either $3x^2 + y^2 = p$, or else, $x^2 + 3z^2 = p$; in either case, p is of the form $a^2 + 3b^2$.

We have proved that p is expressible as $p = a^2 + 3b^2$ if and only if (*) holds. Since

$$3^{(p-1)/2} \equiv (-1)^{[(p+1)/6]} \pmod p,$$

we see that (*) holds if and only if $(p-1)/2 + [(p+1)/6]$ is even.

If p is a prime greater than 3, then it is of the form $6k + 1$ or $6k - 1$. If $p = 6k + 1$, then $(p-1)/2 + [(p+1)/6] = 3k + k$, and if $p = 6k - 1$, then $(p-1)/2 + [(p+1)/6] = 3k - 1 + k$. Thus, $(p-1)/2 + [(p+1)/6]$ is even when $p = 6k + 1$ and is odd when $p = 6k - 1$. This proves (a).

(b) From the previous problems and (a), we see that p can be expressed in these three ways if and only if

$$p \equiv 1 \pmod 4, \quad p \equiv 1 \text{ or } 3 \pmod 8 \quad \text{and} \quad p \equiv 1 \pmod 6.$$

Equivalently, $p \equiv 1 \pmod 8$ and $p \equiv 1 \pmod 6$, that is, $p - 1$ is divisible by 24. This gives (b). Note that 73 is such a prime, and

$$73 = 3^2 + 8^2 = 1^2 + 2(6^2) = 5^2 + 3(4^2).$$

6. Prove that there exist an infinite number of primes
   (a) of the form $8k + 3$, and
   (b) of the form $6k + 1$.

Solution (a) Assume that there are only a finite number of primes of the form $8k + 3$. Let a be the product of all such primes, and $n = 2a^2 + 1$. Let p be a prime divisor of n. Then, p is relatively prime to a; so p cannot be of the form $8k + 3$ as all such primes are factors of a. Further,

$$-2a^2 \equiv 1 \pmod p, \quad (-2)^{(p-1)/2} a^{p-1} \equiv 1 \pmod p,$$

and $a^{p-1} \equiv 1 \pmod p$. It follows that

$$(-2)^{(p-1)/2} \equiv 1 \pmod p.$$

This implies, as before, that p is of the form $8k + 1$ or $8k + 3$. But p is not of the form $8k + 3$. This shows that every prime divisor of n is of the form $8k + 1$. The product of any number of primes of the form $8k + 1$ is again of the same form. Hence, n must be of the form $8k + 1$. It is clear that a is odd, and so $a^2 \equiv 1 \pmod 8$. Thus, $n \equiv 3 \pmod 8$. This contradiction proves (a).

(b) Assume, as above, that there are only a finite number of primes of the form $6k + 1$. Let a be the product of all such primes, and $n = 12a^2 + 1$. Let p be a prime divisor of n. Then, p is relatively prime to a; so p cannot be of the form $6k + 1$ as all such primes are factors of a. Further,

$$-12a^2 \equiv 1 \pmod p, \quad (-3)^{(p-1)/2}(2a)^{p-1} \equiv 1 \pmod p,$$

and $(2a)^{p-1} \equiv 1 \pmod p$. It follows that

$$(-3)^{(p-1)/2} \equiv 1 \pmod p,$$

which implies, as above, that p is of the form $6k + 1$ and shows that every prime divisor of n is of the form $6k + 1$. This is not possible as all such primes are factors of a. The contradiction proves (b).

7. Prove the following generalisation of Euler's theorem:

$$a^n \equiv a^{n-\phi(n)} \pmod n$$

for all integers a.

Solution If a is relatively prime to n, then $a^{\phi(n)} \equiv 1 \pmod n$, by Euler's theorem, and the result holds in that case. Suppose that $d = \gcd(a, n) > 1$. Let $p_1, \ldots, p_r$ be the distinct prime divisors of d. Then these primes divide a and n. Let $n = p_1^{n_1} \cdots p_r^{n_r} m$, where m is not divisible by any of the primes p_i, $1 \le i \le r$, and $k = n/m$. Then, k and a are relatively prime to m. Thus, $\phi(n) = \phi(m)\phi(k)$, and

$$a^{\phi(m)} \equiv 1 \pmod m, \quad a^{\phi(n)} \equiv 1 \pmod m.$$

We have $\phi(n) = n \prod_{p|n} (1 - 1/p)$, where the product is extended over the distinct prime divisors of n. Since p_i is one of the prime divisors of n, we see that $\phi(n) \le n\,(1 - 1/p_i)$ for $1 \le i \le r$, and hence,

$$n - \phi(n) \ge \frac{n}{p_i} \ge p_i^{n_i - 1} \ge 2^{n_i - 1} \ge n_i.$$

Hence, $a^{n-\phi(n)}$ is divisible by a^{n_i}. Since a is divisible by p_i, this implies that $a^{n-\phi(n)}$ is divisible by $p_i^{n_i}$. This is true for $1 \le i \le r$.

It follows that $a^{n-\phi(n)}$ is divisible by $p_1^{n_1} \cdots p_r^{n_r} = k$. Since $a^{\phi(n)} \equiv 1 \pmod{m}$, we see that

$$a^n - a^{n-\phi(n)} = a^{n-\phi(n)} \left\{ a^{\phi(n)} - 1 \right\}$$

is divisible by $km = n$. This proves the result.

8. Let n be an integer greater than 1. Prove that n is prime if and only if there is a positive integer a such that

$$a^{n-1} \equiv 1 \pmod{n} \quad \text{but} \quad a^d \not\equiv 1 \pmod{n}$$

for every proper divisor d of $n-1$.

Solution If n is prime, then n has a primitive root a; and this a has the above property.

Conversely, suppose that there exists a with this property. If $a^{n-1} - 1$ is divisible by n, then a must necessarily be prime to n. So $a^{\phi(n)} \equiv 1 \pmod{n}$. Let h be the smallest positive integer such that $a^h \equiv 1 \pmod{n}$. Then, $h|\phi(n)$ and $h|n-1$. Hence, $h \le \phi(n) \le n-1$. It is given that

$$h < n-1 \quad \text{and} \quad h|n-1 \Longrightarrow a^h \not\equiv 1 \pmod{n}.$$

So we must have $h = n-1$ which implies that $\phi(n) = n-1$. This is possible only when n is prime.

9. (a) Let p be a prime. If $\gcd(a,p) = 1 = \gcd(n, p-1)$, prove that the congruence $x^n \equiv a \pmod{p}$ has a solution, unique modulo p.

   (b) Let p be a prime of the form $3k+2$. Show that the congruence $x^3 \equiv a \pmod{p}$ has a solution for $1 \le a \le p-1$. It also show that if p is of the form $3k+1$, then this congruence has a solution for only one-third of such values of a.

Solution (a) Since n is relatively prime to $p-1$ there exist positive integers r and s such that $nr - (p-1)s = 1$. Let $u^n \equiv a \pmod{p}$ with $1 \le u \le p-1$. Now, $u^{(p-1)s} \equiv 1 \pmod{p}$ since $u^{p-1} \equiv 1 \pmod{p}$. Thus,

$$u^{nr} = u^{1+(p-1)s} \equiv u \pmod{p}.$$

Since $u^{nr} \equiv a^r \pmod{p}$, we get $u \equiv a^r \pmod{p}$. Similarly, if $v^n \equiv a \pmod{p}$, $1 \le v \le p-1$, then $v \equiv a^r \pmod{p}$, and hence $v = u$. This, first of all shows that if the congruence $x^n \equiv a$

(mod p) has a solution, then it is unique modulo p. It also shows that the numbers $1^n, 2^n, \ldots, (p-1)^n$ are incongruent modulo p; each of them is relatively prime to p. It follows that if a is relatively prime to p, then $a \equiv u^n \pmod{p}$ for some u with $1 \leq u \leq p-1$.

(b) If $p = 3k+2$, then $\gcd(3, p-1) = 1$, and so the first statement follows from (a).

Let $p = 3k+1$, and g be a primitive root of p. Then, $g^{p-1} \equiv 1 \pmod{p}$ while $g^h \not\equiv 1 \pmod{p}$ for $1 \leq h < p-1 = 3k$. If $a = g^{3i}$, $1 \leq i \leq k$, then the congruence $x^3 \equiv a \pmod{p}$ has a solution $x = g^i$. Thus, we get at least k incongruent values of a for which the congruence has a solution. Moreover, if $u^3 \equiv a \pmod{p}$, $v = ug^k$ and $w = ug^{2k}$, then u, v, w are three incongruent solutions of this. Since the maximum number of integers which are incongruent modulo p is $p-1 = 3k$, we get the result.

10. (a) Let $n \geq 3$ and k be odd. Prove that the congruence $x^2 \equiv k \pmod{2^n}$ has a solution if and only if $k \equiv 1 \pmod{8}$.

(b) Let $n \geq 3$ and $k \equiv 1 \pmod{8}$. Prove that the congruence $x^2 \equiv k \pmod{2^n}$ has exactly four incongruent solutions.

Solution (a) Suppose that there is an integer u such that $u^2 \equiv k \pmod{2^n}$. Then, $u^2 \equiv k \pmod{8}$ as $n \geq 3$. Since k is odd, this implies that u is odd and hence, $k \equiv u^2 \equiv 1 \pmod{8}$.

For the converse, let $k \equiv 1 \pmod{8}$. We have to prove that if $n \geq 3$, then the congruence $x^2 \equiv k \pmod{2^n}$ has a solution. If $n = 3$, then $x = 1$ is a solution.

Assume that the result holds for some $n \geq 3$. Then there is u such that $u^2 \equiv k \pmod{2^n}$, that is, $u^2 = k + 2^n m$ for some m. If m is even, then u satisfies the congruence $x^2 \equiv k \pmod{2^{n+1}}$. If m is odd, let $v = u + 2^{n-1}$. Then,

$$v^2 = u^2 + 2^n u + 2^{2n-2} = k + 2^n(m+u) + 2^{2n-2} \equiv k \pmod{2^{n+1}},$$

as u is odd and hence, $m+u$ is even. Thus, in either case, the congruence $x^2 \equiv k \pmod{2^{n+1}}$ has a solution. This proves the result by induction.

(b) By (a), the congruence $x^2 \equiv k \pmod{2^n}$ has at least one solution. Let u and v be solutions. They are odd, and so $v+u$ and $v-u$ are even. Further,

$$\frac{v+u}{2}\frac{v-u}{2} = \frac{v^2-u^2}{4} \equiv 0 \pmod{2^{n-2}}.$$

The sum of $(v+u)/2$ and $(v-u)/2$ is v, which is odd. So one of them must be odd and the other even. First suppose that $(v+u)/2$ is odd. Then,

$$\frac{v-u}{2} \equiv 0 \pmod{2^{n-2}}, \quad v-u \equiv 0 \pmod{2^{n-1}}.$$

Hence, $v-u \equiv 0$ or $2^{n-1} \pmod{2^n}$. Let $u_1 = u + 2^{n-1}$. Then, v is congruent modulo 2^n to u or u_1. Further, u and u_1 are incongruent modulo 2^n. Similarly, if $(v-u)/2$ is odd, then v is congruent modulo 2^n to $u_2 = -u$ or $u_3 = -u + 2^{n-1}$. Further, u_2 and u_3 are incongruent modulo 2^n.

We proved that if u is a solution of $x^2 \equiv k \pmod{2^n}$, then any other solution v is congruent to u, u_1, u_2 or u_3. Since u is odd, we see that $2u$ is not divisible by 4. So none of

$$u-u_2 = u_1-u_3 = 2u, \quad u-u_3 = 2u-2^{n-1}, \quad u_1-u_2 = 2u+2^{n-1}$$

is divisible by 4. It follows that the four solutions u, u_1, u_2, u_3 are incongruent modulo 2^n. Hence there are exactly four incongruent solutions.

11. Let d be a positive integer which is not a perfect square. It is known that, for any such d, the equation $x^2 - dy^2 = 1$ has a solution in positive integers x, y.

(a) Prove that $x^2 - dy^2 = -1$ has no solution if $d \equiv 3 \pmod 4$.

(b) Prove that if $x^2 - dy^2 = N$ has one solution, it has infinitely many solutions.

(c) If k is any positive integer, prove that there are infinitely many solutions in integers of $x^2 - dy^2 = 1$ with $k|y$.

Solution (a) If $d \equiv 3 \pmod 4$, then $x^2 - dy^2 \equiv x^2 + y^2 \pmod 4$. But $m^2 \equiv 0$ or $1 \pmod 4$ for any integer m. So $x^2 + y^2 \not\equiv -1 \pmod 4$. Hence, the equation cannot have a solution.

(b) We use the fact that there exist positive integers x_1, y_1 such that $x_1^2 - dy_1^2 = 1$. If x and y are positive integers such that $x^2 - dy^2 = N$, let

$$u = xx_1 + dyy_1, \quad v = xy_1 + yx_1.$$

Then,

$$u^2 - dv^2 = (x^2 - dy^2)(x_1^2 - dy_1^2) = N.$$

Thus, one solution (x, y) gives another solution (u, v) with $u > x$, $v > y$. This can be repeated any number of times.

(c) If d is not a perfect square, then $k^2 d$ is not a perfect square. So we can find positive integers u, v such that $u^2 - k^2 dv^2 = 1$. From one solution, we can form any number of solutions, by (b). If $x = u$ and $y = kv$, then $x^2 - dy^2 = 1$ and $k|y$.

12. Determine all the Pythagorean triples (x, y, z) forming (a) an arithmetic progression, (b) a geometric progression.

Solution (a) Suppose that x, y, z is a Pythagorean triple forming an arithmetic progression. If $d = y - x$, then $x = y - d$, $z = y + d$, and so

$$(y-d)^2 + y^2 = (y+d)^2, \quad y^2 - 2yd + d^2 + y^2 = y^2 + 2yd + d^2.$$

Thus, $y^2 = 4yd$, $y = 4d$. Hence, $x = 3d$ and $z = 5d$. A Pythagorean triple forming an arithmetic progression is $3d, 4d, 5d$ for some positive integer d.

(b) Suppose that x, y, z is a Pythagorean triple forming a geometric progression. Then, $y^2 = zx$, and $x = kx_1$, $y = ky_1$, $z = kz_1$, where x_1, y_1, z_1 is primitive. Thus, $y_1^2 = z_1 x_1$, and

$$z_1^2 = x_1^2 + y_1^2 = x_1^2 + z_1 x_1 = x_1(x_1 + z_1).$$

Since z_1 is relatively prime to x_1, this implies that z_1 divides the other factor $x_1 + z_1$, which in turn implies that $z_1|x_1$. This contradiction shows that there is no Pythagorean triple which can form a geometric progression.

13. If x, y, z is a primitive Pythagorean triple, prove that $x + y$ and $x - y$ are congruent modulo 8 to 1 or -1.

Solution There exist positive integers r, s such that

$$x = r^2 - s^2,\ y = 2rs, \quad \text{or else,} \quad x = 2rs,\ y = r^2 - s^2,$$

where $\gcd(r, s) = 1$ and $r \not\equiv s \pmod 2$. Thus,

$$x + y = r^2 - s^2 + 2rs, \quad x - y = \pm(r^2 - s^2 - 2rs).$$

First suppose that r is odd. Then, s is even and $r \pm s$ is odd. Hence, $(r \pm s)^2 \equiv 1 \pmod 8$ and $2s^2 \equiv 0 \pmod 8$. Since,

$$r^2 - s^2 + 2rs = (r+s)^2 - 2s^2,$$
$$r^2 - s^2 - 2rs = (r-s)^2 - 2s^2,$$

we see that $x + y \equiv 1 \pmod 8$ and $x - y \equiv 1$ or $-1 \pmod 8$. On the other hand, if r is even and s is odd, then $(r \pm s)^2 \equiv 1 \pmod 8$, again, $2r^2 \equiv 0 \pmod 8$, and

$$r^2 - s^2 + 2rs = 2r^2 - (r - s)^2, \quad r^2 - s^2 - 2rs = 2r^2 - (r + s)^2.$$

This shows that $x + y \equiv -1 \pmod 8$ and $x - y \equiv -1$ or $1 \pmod 8$.

14. (a) Prove that every positive integer n can be expressed in the form $n = x^2 + y^2 - z^2$.

(b) Prove that if x, y, z are integers such that $x^2 + y^2 = z^2$, then xy is divisible by 12, and xyz is divisible by 60.

(c) Prove that if x, y, z are non-zero integers such that $\dfrac{1}{x^2} + \dfrac{1}{y^2} = \dfrac{1}{z^2}$, then xy is divisible by 60.

Solution (a) If n is odd, then

$$\left(\frac{n+1}{2}\right)^2 + 0^2 - \left(\frac{n-1}{2}\right)^2 = n,$$

and if n even, then

$$\left(\frac{n}{2}\right)^2 + 1^2 - \left(\frac{n}{2} - 1\right)^2 = n,$$

This proves (a).

(b) If $x^2 + y^2 = z^2$, then $x = kx_1$, $y = ky_1$, $z = kz_1$, where k is a positive integer and x_1, y_1, z_1 is a primitive Pythagorean triple. Thus,

$$x_1 = r^2 - s^2,\ y_1 = 2rs, \quad \text{or else,} \quad x_1 = 2rs,\ y_1 = r^2 - s^2,$$

and $z_1 = r^2 + s^2$, where $\gcd(r, s) = 1$ and $r \not\equiv s \pmod 2$. In either case,

$$x_1 y_1 = (r^2 - s^2)(2rs), \quad x_1 y_1 z_1 = (r^4 - s^4)(2rs).$$

Now, $2rs$ is divisible by 4 since r or s is even. If neither r nor s is divisible by 3, then $r^2 \equiv 1 \equiv s^2 \pmod 3$, in which case, $r^2 - s^2$ is divisible by 3. Hence, rs or $r^2 - s^2$ is divisible by 3. Thus, $x_1 y_1$ is always divisible by 3 as well as 4, and so by 12.

Similarly, if neither r nor s is divisible by 5, then $r^4 \equiv 1 \equiv s^4 \pmod 5$, in which case, $r^4 - s^4$ is divisible by 5. Hence, rs or $r^4 - s^4$

is divisible by 5. Thus, $x_1y_1z_1$ is always divisible by 3, 4 and 5, and so by 60. This completes the proof of (b) since $xy = k^2x_1y_1$ and $xyz = k^3x_1y_1z_1$.

(c) If x, y, z are as given, then $y^2 + x^2 = (xy/z)^2$. Thus, xy/z is a rational number whose square is an integer. Hence, $z_1 = xy/z$ is an integer, and x, y, z_1 is a Pythagorean triple. It follows from (b) that xy is divisible by 12, and xyz_1 is divisible by 60. Since $xyz_1 = (xy)^2/z$, we see that 5 divides xy. Consequently, xy is divisible by 12 and 5, and hence, by 60.

15. Consider the collection of primitive Pythagorean triples x, y, z, where x, z are odd and y is even. Prove that there exist infinitely many such triples (x, y, z) in which

(a) the even member y is a perfect square,

(b) the odd member x is a perfect square,

(c) the odd member z is a perfect square.

Solution Let m be any positive integer. If $r = 2m^2$ and $s = 1$, then r is even, s is odd and $\gcd(r, s) = 1$. Hence, $x = r^2 - s^2$, $y = 2rs$, $z = r^2 + s^2$ gives a primitive Pythagorean triple (x, y, z) in which $y = (2m)^2$. This proves (a).

If $s = 2m(m + 1)$ and $r = s + 1$, then s is even, r is odd and $\gcd(r, s) = 1$. Hence, $x = r^2 - s^2$, $y = 2rs$, $z = r^2 + s^2$ gives a primitive Pythagorean triple (x, y, z) in which

$$x = (r - s)(r + s) = 2s + 1 = 4m(m + 1) + 1 = (2m + 1)^2.$$

This proves (b).

Finally, let u, v, w be any primitive Pythagorean triple with $u > v$. Then, $\gcd(u, v) = 1$ and $u \not\equiv v \pmod 2$. Let

$$x = u^2 - v^2, \quad y = 2uv, \quad z = u^2 + v^2.$$

Then x, y, z is a primitive Pythagorean triple, and $z = w^2$. This proves (c).

16. Prove that there is no Pythagorean triple (x, y, z) in which both x and y are perfect squares.

Solution Assume that there is a Pythagorean triple (x, y, z) of positive integers in which x and y are perfect squares. Choose one such triple for which z has the least possible value. We shall arrive at a contradiction. Let $x = a^2$ and $y = b^2$.

Let $\gcd(a, b) = d$. Then,

$$\left(\frac{a}{d}\right)^4 + \left(\frac{b}{d}\right)^4 = \left(\frac{z}{d^2}\right)^2.$$

This shows that z/d^2 is a rational number whose square is an integer; so z/d^2 is an integer. Thus we get another triple $\left((a/d)^2, (b/d)^2, z/d^2\right)$ of the same type, with z replaced by z/d^2. The minimality of z implies that $d = 1$. Thus, $\gcd(a, b) = 1$, and hence $\gcd(x, y) = \gcd(a^2, b^2) = 1$. Thus, the Pythagorean triple (x, y, z) is primitive. We may assume that y is even, and x, z are odd. Then there exist positive integers r, s with $\gcd(r, s) = 1$, $r \not\equiv s \pmod 2$ and such that

$$x = r^2 - s^2, \quad y = 2rs, \quad z = r^2 + s^2.$$

If r is even and s odd, then $r^2 - s^2 \equiv -1 \pmod 4$. This is not possible since $r^2 - s^2 = x$ is a perfect square. Hence, r is odd and s is even. Thus, $r(s/2) = y/4$, a perfect square. Since $\gcd(r, s/2) = 1$, this implies that r and $s/2$ are perfect squares.

Now, $r^2 = x + s^2 = a^2 + s^2$. So a, s, r is a Pythagorean triple. This is primitive since $\gcd(r, s) = 1$. Since s is even, we see that

$$a = r_1^2 - s_1^2, \quad s = 2r_1s_1, \quad r = r_1^2 + s_1^2,$$

where r_1, s_1 are positive integers with $\gcd(r_1, s_1) = 1$. Thus, $r_1s_1 = s/2$, a perfect square. Hence, $r_1 = u^2$ and $s_1 = v^2$, where u, v are positive integers. But r is also a perfect square. If $r = w^2$, then

$$u^4 + v^4 = r_1^2 + s_1^2 = r = w^2.$$

Thus, u^2, v^2, w is a triple of the same type as we considered in the beginning. But $z = r^2 + s^2 > r^2 = w^4$. Hence $z > w$, giving a contradiction to the minimality of z. This completes the proof.

17. Let (x, y, z) be a primitive Pythagorean triple in which z is a perfect square. Prove that neither x nor y can be a perfect square.

Solution Let (x, y, z) be a primitive Pythagorean triple in which y is even, and z is a perfect square.

(i) We shall first prove that y cannot be a perfect square. Assume that there exist primitive Pythagorean triples (x, y, z) in which y is even, and y and z are perfect squares. Choose one such triple for which z has the least possible value. There exist positive integers r, s such that $\gcd(r, s) = 1$, r is odd, s is even, and also such that

$$x = |r^2 - s^2|, \quad y = 2rs, \quad z = r^2 + s^2.$$

Then, $r(s/2) = y/4$, a perfect square. Since $\gcd(r, s) = 1$, this implies that r and $s/2$ are perfect squares. If $z = a^2$, then $r^2 + s^2 = z = a^2$. Thus, (r, s, a) is a Pythagorean triple, which is primitive and in which s is even. Hence, there exist r_1, s_1 with $\gcd(r_1, s_1) = 1$, $r_1 \not\equiv s_1 \pmod 2$, and also such that

$$r = r_1^2 - s_1^2, \quad s = 2r_1s_1, \quad a = r_1^2 + s_1^2.$$

If r_1 is even, then $r \equiv -1 \pmod 4$, which is not possible since r is a square. So r_1 is odd and s_1 is even. Further, $r_1s_1 = s/2$, a perfect square; hence, r_1 and s_1 are perfect squares. If $r = u^2$, we get $r_1^2 = u^2 + s_1^2$. Thus, we get a Pythagorean triple (u, s_1, r_1) which is primitive since $\gcd(r_1, s_1) = 1$. Since r_1 and s_1 are perfect squares and s_1 is even, it is of the same type as (x, y, z), which we considered in the beginning. Further, $z = a^2 > r_1^4 \geq r_1$. This contradicts the minimality of z and proves the desired result in the present case.

(ii) Next suppose that x is a perfect square. Let $(z - x)/2 = v$ and $(z + x)/2 = w$. Since z and x are squares of odd integers, we see that $zx = u^2$ for some odd positive integer u, and that v and w are positive integers of which v is even and w is odd. Further, $\gcd(v, w) = \gcd(z, x) = 1$, and $vw = (z^2 - x^2)/4 = y^2/4$, a perfect square. Hence, v and w are perfect squares. Moreover, $u^2 + v^2 = w^2$. By (i), this is not possible since v and w are perfect squares. This completes the proof.

18. Consider the collection of primitive Pythagorean triples (x, y, z) where x, z are odd and y is even. Prove that, for a given n, there exist at least n such triples having

    (a) the same value for x,

    (b) the same value for y.

Solution (a) Let $p_1, \ldots, p_n$ be n distinct odd primes, each of which is congruent to 1 modulo 4, and let N be their product. For $1 \leq i \leq n$, let

$$u_i = \frac{N}{p_i}, \quad r_i = \frac{1}{2}(u_i + p_i), \quad s_i = \frac{1}{2}(u_i - p_i).$$

Then, $\gcd(u_i, p_i) = 1$ and $N \equiv 1 \equiv u_i \pmod 4$. Moreover,

$$2r_i = u_i + p_i \equiv 2 \pmod 4, \quad 2s_i = u_i - p_i \equiv 0 \pmod 4.$$

Hence, r_i is an odd integer and s_i is an even integer. Every common divisor of r_i and s_i will divide their sum u_i and difference p_i; so r_i is relatively prime to s_i. Let

$$x_i = r_i^2 - s_i^2, \quad y_i = 2r_i|s_i|, \quad z_i = r_i^2 + s_i^2.$$

This gives n primitive Pythagorean triples (x_i, y_i, z_i), $1 \le i \le n$. Now,

$$x_i = (r_i + s_i)(r_i - s_i) = u_i p_i = N.$$

Thus, the odd member x_i has the same value N in these triples.

(b) Let $p_1, \ldots, p_n$ be n distinct odd primes, and N be their product and $u_i = N/p_i$, as before. Let $r_i = 2u_i$ and $s_i = p_i$. Then r_i is an even integer, s_i is an odd integer, and $(r_i, s_i) = 1$. Defining x_i, y_i, z_i as before, we get n primitive Pythagorean triples. The even member $y_i = 2u_i p_i = 2N$ has the same value for all i.

19. Prove that there exits an infinite number of primitive Pythagorean triples (x, y, z) in which x and y are consecutive positive integers. Give three such triples.

Solution First note that if (x, y, z) is a Pythagorean triple in which x and y are consecutive positive integers, then $\gcd(x, y) = 1$, and so (x, y, z) is primitive. Suppose that $(x, x+1, z)$ is a Pythagorean triple where x, z are odd. Thus, $2x^2 + 2x + 1 = z^2$. Let

$$u = 3x + 2z + 1, \quad w = 4x + 3z + 2.$$

Then, $u - x = 2x + 2z + 1$ and $u + x = 4x + 2z + 1$. Thus,

$$w + z = 4x + 4z + 2 = 2(u - x), \quad w - z = 4x + 2z + 2 = u + x + 1.$$

It follows that,

$$\begin{aligned} w^2 - z^2 &= 2(u - x)(u + x + 1) = 2(u^2 - x^2 + u - x) \\ &= (2u^2 + 2u + 1) - (2x^2 + 2x + 1). \end{aligned}$$

Since $2x^2 + 2x + 1 = z^2$, this implies that $w^2 = 2u^2 + 2u + 1 = u^2 + (u+1)^2$. Thus, $(u, u+1, w)$ is a Pythagorean triple of the desired type, and is different from $(x, x+1, z)$ since $u > x$. Starting from the triple $(3, 4, 5)$ we can find any number of such triples. Thus, $(3, 4, 5)$ gives $(20, 21, 29)$ which in turn yields $(119, 120, 169)$.

20. (a) Prove that there exist infinitely many primitive Pythagorean triples (x, y, z) in which x and y are consecutive triangular numbers.

    (b) Deduce from (a) that there exist infinitely many triangular numbers that are perfect squares. Obtain three such triangular numbers.

Solution By the previous problem, there exist an infinite number of Pythagorean triples of the form $(x, x+1, z)$. Choose one such.

(a) Let $n = 2x$, and $t_n = n(n+1)/2$ be the nth triangular number. Then, $t_n = x(2x+1)$ and $t_{n+1} = (2x+1)(x+1)$. Hence,

$$\begin{aligned} t_n^2 + t_{n+1}^2 &= x^2(2x+1)^2 + (2x+1)^2(x+1)^2 \\ &= (2x+1)^2\{x^2 + (x+1)^2\} = (2x+1)^2 z^2. \end{aligned}$$

Thus, $t_n, t_{n+1}, (n+1)z$, where $n = 2x$, forms a Pythagorean triple.

(b) Let $n = z - x - 1$ and $u = z - 2x - 1$. Then,

$$u^2 = z^2 + 4x^2 + 1 - 4zx - 2z + 4x.$$

Since $z^2 = 2x^2 + 2x + 1$, we can write $4x^2 + 4x + 1$ as $z^2 + 2x^2 + 2x$. Hence,

$$\begin{aligned} u^2 &= 2z^2 - 4zx + 2x^2 + 2x - 2z = 2\{(z-x)^2 - (z-x)\} \\ &= 2(z-x)(z-x-1) = 2(n+1)n. \end{aligned}$$

Hence, $u^2 = 4t_n$. Thus, u is even, and $t_n = (u/2)^2$, a perfect square.

In the previous problem we found three Pythagorean triples $(3, 4, 5)$, $(20, 21, 29)$ and $(119, 120, 169)$, where the first two members are consecutive. The triple $(3, 4, 5)$ corresponds to $n = 1$, $t_n = 1$. The triple $(20, 21, 29)$ corresponds to $n = 8$, $t_n = 36$, and $(119, 120, 169)$ corresponds to $n = 49$, $t_n = 35^2$.

21. Prove that we can find a sequence $a_1, a_2, a_3, \ldots$ of positive integers such that, for all $n \geq 2$,

$$a_1^2 + a_2^2 + \cdots + a_n^2 = \text{the square of an integer.}$$

Solution Let $a_1 = 3$ and $a_2 = 4$. Then, $a_1^2 + a_2^2 = 5^2$. Assume that

$$a_1^2 + a_2^2 + \cdots + a_n^2 = N^2,$$

for some odd integer N. This is true when $n = 2$. Let

$$a_{n+1} = \frac{N^2 - 1}{2} \quad \text{and} \quad N_1 = \frac{N^2 + 1}{2}.$$

Since $N^2 - 1$ is divisible by 8, we see that a_{n+1} is an even integer, and N_1 is odd integer. Further,

$$N_1^2 = \left(\frac{N^2 - 1}{2}\right)^2 + N^2 = a_{n+1}^2 + (a_1^2 + a_2^2 + \cdots + a_n^2).$$

This proves the result by induction on n.

22. Prove that the radius of the incircle of a Pythagorean triangle is always an integer.

Solution A Pythagorean triangle is a right-angled triangle with integer sides. The radius r of the incircle is given by the formula $r = \Delta/s$, where Δ is the area and s is the semi-perimeter. Let x, y, z be the sides where z is the hypotenuse. Now,

$$(x + y + z)(x + y - z) = (x + y)^2 - z^2 = 2xy,$$

This first of all shows that $x + y + z$ or $x + y - z$ is even. But if one of them is even, then the other is also even as they differ by $2z$. Thus, $x + y - z$ is even. Moreover, since $\Delta = xy/2$ and $s = (x + y + z)/2$, we get

$$r = \frac{2\Delta}{2s} = \frac{xy}{x + y + z} = \frac{x + y - z}{2},$$

an integer.

23. For a given positive integer n, show that there is a Pythagorean triangle the radius of whose incircle is n. Show also that if n is an odd prime, then there are precisely three Pythagorean triangles with this property.

Solution Let x, y, z be the sides of the Pythagorean triangle. Suppose that $x < y < z$ so that z is the hypotenuse. Then, as in the previous problem, the radius of the incircle is $r = (x + y - z)/2$. Hence, $r = n$ if and only if $x + y - 2n = z$, that is,

$$z^2 = x^2 + y^2 + 4n^2 + 2xy - 4nx - 4ny.$$

Since $x^2 + y^2 = z^2$, this holds if and only if

$$2n^2 + xy - 2nx - 2ny = 0,$$

that is, $(x - 2n)(y - 2n) = 2n^2$. We can always find x, y, z satisfying these conditions. For example,

$$x = 2n + 1, \quad y = 2n + 2n^2, \quad z = x + y - 2n = 2n^2 + 2n + 1$$

give a Pythagorean triangle, the radius of whose incircle is n. This proves the first part.

The second part: We proved above that the radius of the incircle is n if and only if

$$(x - 2n)(y - 2n) = 2n^2 \quad \text{and} \quad z = x + y - 2n.$$

Since $x + y - 2n = z > y$, we have $x - 2n > 0$. Thus, $x - 2n$ and $y - 2n$ are positive divisors of $2n^2$. Let n be an odd prime. Then, $2n^2$ has exactly 6 positive divisors, namely, 1, 2, n, $2n$, n^2, $2n^2$. Since $x-2n < y-2n$, we must have $x-2n = 1, 2$ or n, and, correspondingly, $y - 2n = 2n^2, n^2, 2n$. Since $z = x + y - 2n$, we have three solutions given by

(i) $x = 2n + 1,\ y = 2n^2 + 2n,\ z = 2n^2 + 2n + 1,$

(ii) $x = 2n + 2,\ y = n^2 + 2n,\ z = n^2 + 2n + 2,$

(iii) $x = 3n,\ y = 4n,\ z = 5n.$

Since $2n + 1 < 2n + 2 < 3n$, the three solutions are different.

24. Find all Pythagorean triangles whose areas are equal to their perimeters.

Solution Let x, y, z be the sides, where z is the hypotenuse. Then its area is $\frac{1}{2}xy$. So let $\frac{1}{2}xy = x + y + z$. Then, $(x + y)^2 - z^2 = 2xy = 4(x + y + z)$. Dividing by $x + y + z$, we get $x + y - z = 4$. As in the previous problem,

$$x^2 + y^2 = z^2 = (x + y - 4)^2 = x^2 + y^2 + 16 + 2xy - 8x - 8y,$$

that is,

$$2xy - 8x - 8y + 16 = 0, \quad (x - 4)(y - 4) = 8.$$

If $x < y$, then $x - 4 = 1, y - 4 = 8$ or $x - 4 = 2, y - 4 = 4$. Thus, we get two triangles with sides $x = 5, y = 12, z = 13$, and $x = 6, y = 8, z = 10$.

25. Prove that

(a) the Diophantine equation $x^4 - y^4 = z^2$ has no solution with $yz \neq 0$,

(b) there are no positive integers a, b, such that both $a^2 + b^2$ and $a^2 - b^2$ are squares of integers, and $a \neq b$,

(c) there is no Pythagorean triangle whose area is the square of an integer.

Solution (a) Suppose that x, y, z are integers with $x^4 - y^4 = z^2$ and $yz \neq 0$. Then $x \neq 0$. Let

$$d = \gcd(x, y), \quad x_1 = \frac{x}{d}, \quad y_1 = \frac{y}{d}, \quad z_1 = \frac{z}{d}.$$

Then, $\gcd(x_1, y_1) = 1$ and $x_1^4 - y_1^4 = z_1^2$. Thus, (z_1, y_1^2, x_1^2) is a Pythagorean triple which is primitive. This is not possible since x_1^2 and y_1^2 are perfect squares [see Problem **??**].

(b) If $a^2 + b^2 = x^2$ and $a^2 - b^2 = y^2$, then $a^4 - b^4 = (xy)^2$. By (a), this is true only when $bxy = 0$, which is not possible if a, b are non-zero and $a \neq b$.

(c) Let x, y, z be the sides of a Pythagorean triangle, where z is the hypotenuse. Suppose that its area is equal to m^2 for some integer m. Then $\frac{1}{2}xy = m^2$, and so

$$z^2 \pm 4m^2 = x^2 + y^2 \pm 2xy = (x \pm y)^2.$$

Thus, $z^2 + (2m)^2$ and $z^2 - (2m)^2$ are perfect squares. If $z = 2m$, then we must have $x = y$ and $z^2 = 2x^2$, which is not possible since $\sqrt{2}$ is irrational. So we get a contradiction by (b), and the proof is complete.

26. Assume, for convenience, that in a primitive Pythagorean triple (x, y, z) of positive integers, the first component x is odd and the second component y is even. Suppose that (x, y, z) and (a, b, c) are two such primitive triples. Let $\gcd(c, z) = 1$ and $w = cz$. Prove that they can be combined to form two triples (u_1, v_1, w) and (u_2, v_2, w) of the same type and having the same third component w.

Solution First observe that $ax \pm by$ is odd, and $ay \pm bx$ is even. Let

$$u_1 = ax + by, \quad v_1 = |ay - bx|,$$
$$u_2 = |ax - by|, \quad v_2 = ay + bx.$$

Then, $u_1^2 + v_1^2 = (a^2 + b^2)(x^2 + y^2) = c^2 z^2$, and, similarly, $u_2^2 + v_2^2 = c^2 z^2$. Thus, (u_1, v_1, w) and (u_2, v_2, w) are Pythagorean triples whose first members are odd and second members are even. It is easily seen that $u_1 \neq u_2$ and $v_1 \neq v_2$. Suppose that $v_1 = 0$. Then $ay = bx$. Since $\gcd(a, b) = 1$, we get $a|x$, and since $\gcd(x, y) = 1$, we get $x|a$. It follows that $x = a$, and hence, $y = b$ and $z = c$. This is not possible since $\gcd(c, z) = 1$. Hence, $v_1 \neq 0$.

To prove that (u_1, v_1, w) and (u_2, v_2, w) are primitive, let

$$d = \gcd(u_1, v_1), \quad d_1 = \gcd(u_2, v_2)$$

We have to prove that $d = d_1 = 1$.

First observe that d divides $u_1^2 + v_1^2 = c^2z^2$. Since $ay - bx = \pm v_1$, and

$$ax(ax + by) - bx(ay - bx) = a^2x^2 + b^2x^2 = c^2x^2,$$

we see that d divides c^2x^2. So, it divides $c^2z^2 - c^2x^2 = c^2y^2$ also. But $\gcd(c^2x^2, c^2y^2) = c^2 \gcd(x^2, y^2) = c^2$. Similarly, since

$$ax(ax + by) + ay(ay - bx) = a^2x^2 + a^2y^2 = a^2z^2,$$

we see that d divides a^2z^2, b^2z^2 and hence, z^2. But $\gcd(c^2, z^2) = 1$. Hence, $d = 1$.

To prove that $d_1 = 1$, note that $ax - by = \pm u_2$. An interchange of x and y in the above calculations will show that $d_1 = 1$.

27. Assume, as before, that in a primitive Pythagorean triple, the first component is odd and the second component is even. Let (x, y, z) be such a triple, p be a prime congruent to 1 modulo 4, $\gcd(z, p) = 1$ and $w = pz$.

(a) Prove that (x, y, z) *generates* two triples (u_1, v_1, w) and (u_2, v_2, w) of the same type and having the same third component w.

(b) Also show that every such triple (u, v, w), with this w, corresponds to a triple (x, y, z) which generates (u, v, w).

Solution (a) It is known that every prime p of the form $4k + 1$ is the sum of two squares. If $p = r^2 + s^2$, let $a = |r^2 - s^2|$ and $b = 2|rs|$. Since p is an odd prime, we see that one of r, s is odd and the other is even, and $\gcd(r, s) = 1$. Consequently, (a, b, p) is a primitive triple, with a odd and b even. Hence, by the previous problem, (x, y, z)

can be combined with (a, b, p) to form two different primitive triples (u_1, v_1, w), (u_2, v_2, w), defined by

$$u_1 = ax + by, \quad v_1 = |ay - bx|,$$
$$u_2 = |ax - by|, \quad v_2 = ay + bx.$$

This proves (a).

(b) Suppose that (u, v, w) and (a, b, p) are as above. Then,

$$\begin{aligned} b^2u^2 &= (p^2 - a^2)(w^2 - v^2) = (p^2 - a^2)(z^2p^2 - v^2) \\ &\equiv a^2v^2 \pmod{p^2}. \end{aligned}$$

Thus, $(bu - av)(bu + av)$ is divisible by p^2. Suppose that both the factors are divisible by p. Then their sum, $2bu$, is divisible by p. But $\gcd(b, p) = 1$ as (a, b, p) is primitive. So $p|u$. Since $\gcd(u, w) = 1$, and $p|w$, we have a contradiction. This shows that p cannot divide both the factors $bu - av$ and $bu + av$. So one of them is divisible by p^2.

If (i) $bu - av \equiv 0 \pmod{p^2}$, let

$$\alpha = au + bv, \quad \beta = |bu - av|,$$

If (ii) $bu + av \equiv 0 \pmod{p^2}$, let

$$\alpha = |au - bv|, \quad \beta = bu + av.$$

Note that α, β are formed out of u, v exactly as (u_1, v_1) and (u_2, v_2) are generated from x, y. Hence,

$$\alpha^2 + \beta^2 = (u^2 + v^2)(a^2 + b^2) = w^2p^2 = p^4z^2.$$

Moreover, $\beta \equiv 0 \pmod{p^2}$ in either case. Let

$$x = \frac{\alpha}{p^2}, \quad y = \frac{\beta}{p^2}.$$

Thus, y is a positive integer, and $x^2 + y^2 = z^2$. So x is also a positive integer; and (x, y, z) is Pythagorean. It is clear that α is odd and β is even, in either case. So x is odd and y is even.

To prove that (x, y, z) is primitive, let $\gcd(x, y) = d$. First consider case (i). We have

$$a(au + bv) + b(bu - av) = (a^2 + b^2)u = p^2u,$$
$$b(au + bv) - a(bu - av) = (b^2 + a^2)v = p^2v,$$

and $bu - av = \pm\beta$. It follows that $u = ax \pm by$, and $v = bx \pm ay$. Similarly, in case (ii) we have

$$a(au - bv) + b(bu + av) = (a^2 + b^2)u = p^2u,$$

$$-b(au - bv) + a(bu + av) = (b^2 + a^2)v = p^2v,$$

and $au - bv = \pm\alpha$. Hence, $u = \pm ax + by$, and $v = \pm bx + ay$. It follows that d divides u and v, in either case. But $\gcd(u, v) = 1$; hence $d = 1$. Thus, (x, y, z) is Pythagorean triple of the desired type.

Suppose that (x, y, z) generates (u_1, v_1, w) and (u_2, v_2, w), where $u_1 = ax + by$ and $u_2 = |ax - by|$. We have found that $u = ax \pm by$ or $u = \pm ax + by$. Hence, $u = ax + by$ or $|ax - by|$. Thus, $u = u_1$ or u_2; so, $v = v_1$ or v_2. This shows that (u, v, w) is the same as (u_1, v_1, w) or (u_2, v_2, w) and completes the proof.

28. Assume, as in the previous problems, that in a primitive Pythagorean triple, the first component is odd and the second component is even. Suppose that there are exactly m such triples (x_i, y_i, z), $i = 1, \ldots, m$, having the same third component z. Let p be a prime of the form $4k+1$ relatively prime to z, and $w = pz$. Prove that there exist exactly $2m$ such primitive triples having the same third component w.

Solution It is known that every prime p of the form $4k + 1$ is the sum of two squares. If $p = r^2 + s^2$, let $a = |r^2 - s^2|$ and $b = 2|rs|$. Since p is an odd prime, we see that one of r, s is odd and the other is even, and $(r, s) = 1$. Consequently, (a, b, p) is a primitive Pythagorean triple, with a being odd and b even. Hence, by the previous problem, each triple (x_i, y_i, z) can be combined with (a, b, p) to form two different primitive Pythagorean triples (u_{i1}, v_{i1}, w), (u_{i2}, v_{i2}, w), defined, for $1 \le i \le m$, as follows:

$$u_{i1} = ax_i + by_i, \quad v_{i1} = |ay_i - bx_i|,$$

$$u_{i2} = |ax_i - by_i|, \quad v_{i2} = ay_i + bx_i.$$

Here u_{i1}, u_{i2} are odd and v_{i1}, v_{i2} are even.

We shall prove that $u_{i1}, u_{i2}, u_{j1}, u_{j2}$ are distinct for all $i, j, i \neq j$. It will then follow that the resulting $2m$ Pythagorean triples are different. Note that $u_{i1} \neq u_{i2}$ and $u_{j1} \neq u_{j2}$.

(i) Suppose that $u_{i1} = u_{j1}$, that is, $ax_i + by_i = ax_j + by_j$,

$$a(x_i - x_j) = b(y_j - y_i). \tag{1}$$

Since $\gcd(a, b) = 1$, this implies that a divides $y_j - y_i$, and so

$$\frac{x_i - x_j}{b} = \frac{y_j - y_i}{a} = k, \text{ an integer.}$$

Hence,

$$\begin{aligned} x_j^2 + y_j^2 &= (x_i - kb)^2 + (y_i + ka)^2 \\ &= x_i^2 + y_i^2 + 2k(ay_i - bx_i) + k^2(a^2 + b^2). \end{aligned}$$

Since $x_j^2 + y_j^2 = z^2 = x_i^2 + y_i^2$, we can cancel these two terms. Since $a^2 + b^2 = p^2$, this gives

$$k^2p^2 = 2|k(ay_i - bx_i)| = 2|k|v_{i1}. \tag{2}$$

Thus, $|k|p^2 = 2v_{i1}$, which implies that $p|v_{i1}$. Since $p|w$, this shows that $\gcd(v_{i1}, w) > 1$. This cannot be true since (u_{i1}, v_{i1}, w) is primitive. It follows that $u_{i1} \neq u_{j1}$.

(ii) Next, suppose that $u_{i1} = u_{j2}$. Then, $ax_i + by_i = |ax_j - by_j|$, that is,

$$ax_i + by_i = ax_j - by_j, \quad \text{or,} \quad ax_i + by_i = -ax_j + by_j.$$

In the first case, $a(x_i - x_j) = b(-y_j - y_i)$. This is the same as (1) with y_j changed to $-y_j$. Since $x_j^2 + (-y_j)^2 = z^2$, the same argument as in (i) gives (2). In the other case, $a(x_i + x_j) = b(y_j - y_i)$. This is the same as (1) with x_j changed to $-x_j$. Since $(-x_j)^2 + y_j^2 = z^2$, we again get (2). Thus, in either case, $p|v_{i1}$, which gives a contradiction as in (i). Hence, $u_{i1} \neq u_{j2}$. Interchanging i and j, we get $u_{j1} \neq u_{i2}$.

(iii) It remains to show that $u_{i2} \neq u_{j2}$. If this is false, $|ax_i - by_i| = |ax_j - by_j|$. Then,

$$ax_i - by_i = ax_j - by_j, \quad \text{or,} \quad ax_i - by_i = -ax_j + by_j.$$

In the first case, $a(x_i - x_j) = -b(y_j - y_i)$. This is the same as (1) with b changed to $-b$. Since $a^2 + (-b)^2 = p^2$, the same argument as in (i) gives

$$k^2p^2 = 2|k(ay_i + bx_i)| = 2|k|v_{i2}.$$

In the other case, $a(x_i + x_j) = b(y_j + y_i)$. This is the same as (1) with x_j changed to $-x_j$ and y_i changed to $-y_i$. Since $(-x_j)^2 + y_j^2 = z^2 =$

$x_i^2+(-y_i)^2$, we again get (2). Thus, we see that $p|v_{i2}$ or $p|v_{i1}$; in either case, we get a contradiction as in (i).

We have proved that there exist at least $2m$ primitive Pythagorean triples of the given type and having the same third component w. It remains to prove that there is no more such triples. Suppose that (u, v, w) is such a triple. Then, (u, v, w) corresponds to a triple (x, y, z) which generates (u, v, w) [Problem ??]. The said (x, y, z) must be the same as some (x_i, y_i, z), and hence, (u, v, w) is the same as (u_{i1}, v_{i1}, w) or (u_{i2}, v_{i2}, w). This completes the proof.

29. Let N be the product of k distinct primes each of which is congruent to 1 modulo 4. Prove that there exist exactly 2^{k-1} primitive Pythagorean triples (x_i, y_i, N) whose first components x_i are odd and second components y_i are even, and have the same third component N.

Solution It is known that if p is prime and $p \equiv 1 \pmod 4$, then $p = r^2+s^2$, where r, s are positive integers. Moreover, such an expression of p is unique except for the order of summands. Let $a = |r^2-s^2|$ and $b = 2rs$. Then (a, b, p) is a Pythagorean triple, which is primitive since $\gcd(r, s) = 1$. The second component b is even. Suppose that (a_1, b_1, p) is another such triple. Then, $a_1 = |r_1^2 - s_1^2|$, $b_1 = 2r_1s_1$ and $p = r_1^2+s_1^2$. Because of the uniqueness stated above, (a_1, b_1, p) is the same (a, b, p). If $k = 1$, then $N = p$ and $2^{k-1} = 1$. So the result is true if $k = 1$.

Use induction on k. Assume that the result is true for $k-1$ primes. Let $N = p_1 \cdots p_k$, where p_i are distinct and are congruent to 1 modulo 4. Then $N = N_1 p_k$. By hypothesis, there are exactly $m = 2^{k-2}$ triples in each of which the first component is odd, second component is even and third component is N_1. So, by the previous problem, there are exactly $2m = 2^{k-1}$ triples of this type with the same third component $N_1 p_k = N$. This proves the result.

30. As in the previous problem. let N be the product of k distinct primes each of which is congruent to 1 modulo 4.

    (a) Determine the number of Pythagorean triples (x, y, N) having the same third component N.

    (b) Find all Pythagorean triples $(x, y, 1105)$.

Solution (a) Let (x, y, N) be a Pythagorean triple. So there is a positive integer k such that $= ka$, $y = kb$ and $N = kN_1$, where (a, b, N_1) is a primitive Pythagorean triple. Then, N_1 is a divisor of N and $N_1 > 1$. Thus, N_1 is a product of some the prime divisors of N.

While counting the number of Pythagorean triples we do not distinguish between (x, y, z) and (y, x, z). So we assume, as in the previous problems, that in a primitive Pythagorean triple, the first component is odd.

Suppose that N_1 is the product of i primes, $1 \leq i \leq k$, each of which is a divisor of N. There are $\binom{k}{i}$ such values for N_1. By the previous problem,there are precisely 2^{i-1} primitive triples (a, b, N_1). If $k = N/N_1$, let $x = ka$ and $y = kb$. Then (x, y, N) is a Pythagorean triple; it is uniquely determined by (a, b, N_1). Hence, there are 2^{i-1} Pythagorean triples determined by a value of N_1. This is true for $1 \leq i \leq k$. It follows that the total number of Pythagorean triples is

$$\binom{k}{1} \cdot 1 + \binom{k}{2} \cdot 2 + \cdots + \binom{k}{k} \cdot 2^{k-1} = \frac{1}{2}\left\{(1+2)^k - 1\right\}.$$

Thus, there are $(3^k - 1)/2$ Pythagorean triples (x, y, N) having the same third component N.

(b) In the same notation as in (a), $N = 1105 = 5 \cdot 13 \cdot 17$. The total number of Pythagorean triples $(x, y, 1105)$ is $(3^3 - 1)/2 = 13$. The possible values for N_1 are 5, 13, 17, 65, 85, 221 and 1105.

If p is prime and $p = r^2 + s^2$, let $a = |r^2 - s^2|$ and $b = 2rs$. This gives a unique triple (a, b, p) in which a is odd and b is even. Since

$$5 = 1^2 + 2^2, \quad 13 = 3^2 + 2^2, \quad 17 = 1^2 + 4^2,$$

we get the primitive triples $(3, 4, 5)$, $(5, 12, 13)$ and $(15, 8, 17)$ corresponding to the primes 5, 13 and 17.

Combining the triples for two primes p_1 and p_2, we get two primitive triples for $N_1 = p_1p_2$. Thus, the triples for 65 are

$$(63, 16, 65) \quad \text{and} \quad (33, 56, 65);$$

those for 85 are

$$(77, 36, 85) \quad \text{and} \quad (13, 84, 85);$$

and those for 221 are

$$(171, 140, 221) \quad \text{and} \quad (21, 220, 221).$$

We have obtained 9 primitive triples whose third components N_1 are proper divisors of 1105. Multiplying by proper values of k, we get 9 Pythagorean triples $(x, y, 1105)$ given by the following 9 pairs (x, y):

$$(663, 884),\ (425, 1020),\ (975, 520),\ (1071, 272),\ (561, 952),$$
$$(1001, 468),\ (169, 1092),\ (855, 700),\ (105, 1100).$$

Combining $(63, 16, 65)$ and $(15, 8, 17)$, we get the triples

$$(1073, 264, 1105), \quad (817, 744, 1105).$$

Combining $(33, 56, 65)$ and $(15, 8, 17)$ we get the triples

$$(943, 576, 1105), \quad (47, 1104, 1105).$$

We have found four more Pythagorean triples $(x, y, 1105)$, making a total of 13.

References

Apostol, Tom M. *Introduction to Analytic Number Theory*. New Delhi: Narosa Publishing House. 1989.

Burton, David M. *Elementary Number Theory*, 6th Edition. Tata McGraw-Hill Publishing Company Limited. 2006.

Hardy G.H. and E.M. Wright. *An Introduction to the Theory of Numbers*. Oxford University Press. 1960.

Niven, Ivan and H.S. Zuckerman. *An Introduction to the Theory of Numbers*. John Wiley and Sons. 1960.

Klamkin, Murray S. *International Mathematical Olympiads, 1978–1985*. Washington DC: The Mathematical Association of America. 1986.

Klamkin, Murray S. *USA Mathematical Olympiads, 1972–1986*. Washington DC: The Mathematical Association of America. 1988.

Pranesachar, C.R., S.A. Shirali, B.J. Venkatachala and C.S. Yogananda. *Mathematical Challenges from Olympiads*. Bangalore: Interline Publishing Pvt. Ltd. 1995.

Greitzer, Samuel L., *International Mathematical Olympiads, 1959–1977*. Washington DC: The Mathematical Association of America.1978.

Index